Introduction to Workplace Safety and Health Management

A Systems Thinking Approach

Third Edition

World Scientific Series on the Built Environment

ISSN: 2737-4831

Series Editor: Willie Chee Keong Tan
National University of Singapore, Singapore)

Published

Vol. 8   *Introduction to Workplace Safety and Health Management:*
         A Systems Thinking Approach
         Third Edition
         by Yang Miang Goh

Vol. 7   *Managing Infrastructure Projects*
         Second Edition
         by Willie Tan

Vol. 6   *Urban Greening Techniques: An Introduction*
         by Chun Liang Tan

Vol. 5   *Urban Management: Managing Cities in Uncertain Times*
         by Willie Tan

Vol. 4   *Facilities Planning and Design: An Introduction for Facility Planners,*
         Facility Project Managers and Facility Managers
         Second Edition
         by Jonathan K M Lian

Vol. 3   *Managing Infrastructure Projects*
         by Willie Tan

Vol. 2   *Introduction to Workplace Safety and Health Management:*
         A Systems Thinking Approach
         Second Edition
         by Yang Miang Goh

Vol. 1   *Contract Administration and Procurement in the Singapore*
         Construction Industry
         Second Edition
         by Pin Lim

World Scientific Series on the Built Environment

Volume 8

Introduction to Workplace Safety and Health Management

A Systems Thinking Approach

Third Edition

Yang Miang Goh

National University of Singapore, Singapore

NEW JERSEY · LONDON · SINGAPORE · BEIJING · SHANGHAI · HONG KONG · TAIPEI · CHENNAI · TOKYO

Published by

World Scientific Publishing Co. Pte. Ltd.

5 Toh Tuck Link, Singapore 596224

USA office: 27 Warren Street, Suite 401-402, Hackensack, NJ 07601

UK office: 57 Shelton Street, Covent Garden, London WC2H 9HE

Library of Congress Control Number: 2024942427

British Library Cataloguing-in-Publication Data
A catalogue record for this book is available from the British Library.

World Scientific Series on the Built Environment — Vol. 8
INTRODUCTION TO WORKPLACE SAFETY AND HEALTH MANAGEMENT
A Systems Thinking Approach
Third Edition

ISBN 978-981-12-9001-5 (hardcover)
ISBN 978-981-12-9006-0 (paperback)
ISBN 978-981-12-9002-2 (ebook for institutions)
ISBN 978-981-12-9003-9 (ebook for individuals)

For any available supplementary material, please visit
https://www.worldscientific.com/worldscibooks/10.1142/13754#t=suppl

Desk Editors: Gregory Lee/Amanda Yun

Typeset by Stallion Press
Email: enquiries@stallionpress.com

About the Author

Goh Yang-Miang is an Associate Professor and Director at the Centre for Project and Facilities Management, Department of The Built Environment (DBE), College of Design and Engineering (CDE), National University of Singapore (NUS).

He holds both a B.Eng. and a Ph.D. in Civil Engineering from NUS. Professor Goh was Assistant Dean of CDE (2023–2024), Deputy Head (Research) (2016–2019) and Deputy Programme Director (B.Sc. Project and Facilities Management) (2013–2016) of DBE, NUS; Assistant Professor at the Department of Building, NUS (2012–2016); Senior Consultant at Det Norske Veritas, Singapore (2011–2012); Senior Lecturer at the School of Public Health, Curtin University, Western Australia (2007–2010); Engineer, Group Leader, Head, and Assistant Director (Investigations) at the Occupational Safety and Health Division at the Ministry of Manpower, Singapore (2005–2007); and Research Engineer/Research Assistant at NUS (2003–2004).

Professor Goh is Deputy Chairman of the Singapore Standards Technical Committee on Safety and Health in Equipment (2024-current); Member of the 2nd and 3rd External Review Panel on SAF Safety (ERPSS) (2017–2023); Member of the Workplace Safety and Health Council Construction and Landscape Committee (2015–2017); and Chairman/Co-Chairman of the Health and Safety Engineering Technical Committee, Institution of Engineers (IES), Singapore (2015–2018). He served as a member of the Panel of Judges for Singapore Land Transport Authority's Annual Safety Award Convention. He previously consulted as an expert witness for WSH-related litigation processes and also provided advice on Myanmar's WSH legislation through the International Labor Organization (ILO).

His research interests are: workplace safety and health, systems thinking, fall protection, safety analytics, machine learning, computer vision, project management, and education technology. Professor Goh has published more than 120 publications, and according to a study by Stanford University, he was recognised as among the top 2% of scientists in the world.

He has also received numerous teaching awards from NUS and a Bronze Asia regional award of the Wharton-QS Reimagine Education Awards (2022).

List of Abbreviations

AARC	Approved Asbestos Removal Contractor
ACC	Approved Crane Contractors
ACoP	Approved Code of Practice
AE	Authorised Examiner
ALARP	As low as reasonably practicable
BBS	Behaviour-based Safety
BCA	Building and Construction Authority
BE	Breakdown event
BIM	Building Information Modelling
BS	British Standards
CE	Contact event
CEO	Chief Executive Officer
CFAC	Contributing factors in accident causation
CHAIR	Construction Hazard Assessment and Implication Review
CLD	Causal Loop Diagram
CSB	Chemical Safety Board
CSQ	Consequence
DfS	Design for Safety
DfSP	DfS Professional
D&B	Design-and-build
ECI	Early Contractor Involvement
ECT	Event Causation Technique
ESG	Environmental, Social and Governance
ETA	Event tree analysis
ETM	Energy Transfer Model
FMEA	Failure mode and effect analysis

FRAM	Functional Resonance Accident Model
FTA	Fault tree analysis
GP Reg	General Provisions Regulations
HAZOP	Hazard and Operability
HFACS	Human Factors Analysis and Classification System
HSE	Health and Safety Executive
ICAM	Incident Cause Analysis Method
IE	Intermediate event
IES	Institution of Engineers Singapore
ILO	International Labor Organization
IR Reg	Incident Reporting Regulations
ISO	International Organization for Standardization
ISOM	Isomerisation
JHA	Job hazard analysis
JPW	Jurong Primewide Pte Ltd
JSA	Job safety analysis
KPI	Key Performance Indicator
LCM	Loss Causation Model
LOPA	Layers of Protection Analysis
LTA	Land Transport Authority
MLCM	Modified Loss Causation Model
MOC	Management of Change
MOM	Ministry of Manpower
MORT	Management Oversight Risk Tree
MOU	Memorandum of Understanding
MTO	Man-Technology-Organisation
NEA	National Environment Agency
NUS	National University of Singapore
OHS	Occupational health and safety
OHSAS	Occupational Health and Safety Management Systems
OIMS	Operations Integrity Management System
OOC Reg	Operations of Cranes Regulations
OSH	Occupational safety and health
OSHA	Occupational Safety and Health Administration
PDCA	Plan-Do-Check-Act

PE	Professional Engineer
PEL	Permissible Exposure Levels
PESTLE	Political, Economic, Social, Technological, Legal, Environmental
PHA	Process hazard analysis
PM	Project Manager
PMD	Personal mobility devices
PPE	Personal protective equipment
PTW	Permit-to-work
RA	Risk Assessment
RACI	Responsible, Accountable, Consulted and Informed
REDAS	Real Estate Developer's Association of Singapore
RFID	Radio frequency identification
RM	Risk Management
RMCP	Risk Management Code of Practice
RPN	Risk priority number
SCAL	Singapore Contractors Association Ltd
SCM	Swiss Cheese Model
SD	Safe Design
SDS	Safety data sheets
SHE	Safety, Health and Environment
SHMS Reg	Safety and Health Management System and Auditing Regulations
SHS	Steel Hollow Section
SLII®	Situational Leadership® II
STAMP	Systems Theoretic Accident Modelling and Process
STOP	Stop, Think, Observe and Plan
SWOT	Strengths, Weaknesses, Opportunities, and Threats
SWP	Safe Work Procedure
VSCC	Vessel Safety Coordination Committee meetings
WAH	Working-at-height
WAH Regs	Workplace Safety and Health (Work at Heights) Regulations

WHS	Work Health and Safety
WSHA	Workplace Safety and Health Act
WSHC	Workplace Safety and Health Council
WSHMS	Workplace Safety and Health Management System
WSHO	Workplace safety and health officer

Contents

Chapter 3 Incident investigation

Chapter 4 Workplace safety and health risk management

Chapter 5 Design for Safety

Chapter 6 Overview of workplace safety and health management systems

Chapter 7 Safety culture and leadership

Chapter 8   Improving safety culture

Chapter 9   Workplace safety and health legislations

Chapter 10   Accident case studies

CHAPTER 1

Introduction

1.1 Importance of workplace safety and health management

Neglecting workplace safety and health[1] (WSH) management can cause accidents and ill health, leading to severe consequences for individuals, families, communities, organisations, and industries. However, as WSH incidents are uncertain events that may not occur even if work is conducted unsafely, many organisations do not see the importance of WSH. Managers may also neglect WSH management when other pressing issues related to revenue, schedule, and client expectations demand their attention and resources.

This book is written for current and future managers overseeing and managing high-risk workplaces like construction worksites, shipyards and factories. Even if managers do not manage operations directly, they make decisions that significantly influence WSH. Thus, this book is also relevant to professionals like project managers, contract managers, engineers, and architects involved in designing, selecting, and planning products and operations. Knowledge of WSH management will also allow managers to identify and select contractors with strong WSH competencies. In addition, with legislations such as the WSH (Design for Safety) Regulations ("DfS Regulations") (Chap. 5), all stakeholders are expected to be proactive in WSH management.

---

[1] Also known as occupational safety and health (OSH) or occupational health and safety (OHS).

Design for Safety (DfS) (also known as prevention through design, safe design, and Construction (Design and Management)) promotes early consideration of safety and health hazards during the design phase of a construction project. Similarly, early safety management during upstream stages has significant benefits in other industries. Early intervention can eliminate or control risks, leading to safer workplaces and processes.

Furthermore, the corporate world is paying more attention to environmental, social and governance (ESG) matters. "Environmental" is about how a company is dealing with sustainability issues. "Social" is about social impact and social good in general. It includes gender diversity, labour rights, and WSH. "Governance" is about the company's decision-making processes and policy framework to ensure transparency and integrity. In many countries, listed companies need to report their ESG performance. Investors can consider the ESG performance metrics when making investment decisions. For example, the Singapore Exchange (2023) requires companies listed on the exchange to report a wide range of ESG indicators, including WSH metrics such as fatalities, injuries, and ill health. The scope of reporting includes employees and "workers who are not employees but whose work and/or workplace is controlled by the organisation" (Singapore Exchange, 2023). These reporting requirements impose pressure on listed companies to have good WSH performance, and they would also require good WSH performance from their partners, suppliers and vendors. Thus, companies, including board members and company directors, need to pay attention to WSH.

1.1.1 Accidents

Accidents happen, and when they do, people suffer. However, many people do not register this simple fact. Even in relatively low-risk environments like universities, there have been incidents of fire, which have caused property damage. For example, the

National University of Singapore (NUS) had fires in August 2012, October 2012, and April 2014. Fortunately, these fires did not result in any significant injury or fatality. However, in 2008, a tower crane collapsed at a worksite in NUS, killing three construction workers.

In Singapore, the Nicoll Highway collapsed on 20 April 2004 (Goh and Soon, 2014), and it remains one of the worst industrial accidents that Singapore has had. The Nicoll Highway collapse led to four deaths, and the body of the foreman, Mr Heng Yeow Peow, who saved several workers during the collapse, was never found. Though the four deaths are very significant, and the impact on the victims' families is immeasurable, it must be remembered that the scale of the collapse of the Nicoll Highway, a busy road linking to the central area of Singapore, could have easily led to more fatalities. After the accident, senior managers of the main contractor, Nishimatsu (partner in the Nishimatsu-Lum Chang Joint Venture), were severely fined (Goh, 2008). The Professional Engineer (PE) cum project coordinator had made severe errors in the design of the diaphragm wall and strut system, which led to the collapse of the cut and cover tunnel. The PE from New Zealand was fined S$160,000 in April 2006 and banned from practice for two years. In addition, Nishimatsu was fined S$200,000, the design manager was fined S$160,000, and the project director was fined S$120,000. The former Land Transport Authority (LTA) project director was fined S$8,000 for failing to exercise due diligence in monitoring excavation works and assessing readings of soil monitoring instruments. The Nicoll Highway station project was delayed by four years, and millions of dollars were spent on reconstructing the highway and adjacent roads. The Nicoll Highway collapse showed that accidents cause human suffering and project failure, and managers, project managers, and executives can be taken to task for not ensuring workplace safety and health.

The Singapore Ministry of Manpower (MOM) has been taking project managers and appointment holders to task for failing to ensure site safety. In one instance, a project coordinator cum lifting supervisor was sentenced to four weeks' imprisonment. Furthermore, the Singapore WSH Council launched the Code of Practice on Chief Executives' and Board of Directors' Workplace Safety and Health Duties in September 2022 (Chap. 7). The Approved Code of Practice (ACoP) is a clear signal to board members and company directors about the importance of WSH. Thus, managers must be competent in and committed to WSH management.

1.1.2 Ill health

Unlike accidents, ill health takes years to develop, and the illness may not be linked to the work the victim conducted many years ago. According to updated International Labor Organization (ILO) estimates (2023), nearly three million workers die annually from work-related accidents and diseases, a 5% increase since 2015. The majority of these deaths, 2.6 million, are due to work-related diseases, while accidents cause an additional 330,000 deaths. The astonishing numbers signal the devastating impact of occupational diseases. The ILO (2023) indicates that circulatory diseases, cancer, and respiratory diseases are the leading causes of work-related fatalities. Other occupational health issues common across industries include noise-induced hearing loss and deafness, dermatitis, musculoskeletal disorders, and occupational asthma.

1.1.3 Environmental pollution

Many WSH issues are related to environmental pollution; for example, polluted air can lead to respiratory and cardiovascular diseases. Environmental pollution can also lead to communities near a workplace experiencing noise and odour originating from the workplace, affecting their quality of life. Workplace pollution also contributes to global sustainability problems such as climate change, loss of biodiversity, and stratospheric ozone depletion. With the global average temperature rising and more extreme and erratic weather occurring in different parts of the world, workplaces must carefully study the environmental aspects (synonymous with hazards in WSH terminology) and impacts (synonymous with accidents or ill health). WSH incidents, such as major fires and structural collapses, can also cause environmental pollution by releasing substances and dust.

Thus, managers need to be familiar with environmental protection concepts and regulations. For example, the Singapore National Environment Agency (NEA) has different Environmental Protection and Management legislations, and workplaces must ensure a systematic environmental management approach. Even though the WSH management concepts and processes discussed in this book are relevant to environmental management, environmental management is not within the scope of this book.

1.1.4 Cost of incidents

WSH incidents are costly. According to Bird *et al.* (2003) (Fig. 1.1), there is another US$6 to US$53 of uninsured costs for each dollar of insured cost. Like a ship captain who only sees the tip of the

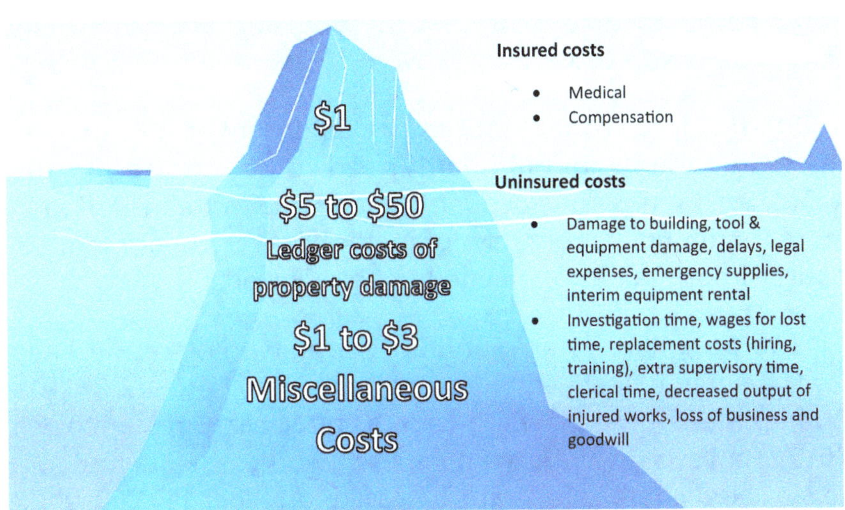

Fig. 1.1   The cost of incidents.

iceberg, many organisations are focused on the insured costs, but they may overlook the massive hidden uninsured costs. The ratio between insured and uninsured costs will vary from incident to incident. However, the key message is that incidents are costly, and it is worthwhile to invest in incident prevention.

1.2 Workplace safety and health statistics

WSH statistics help us understand how different industries perform regarding WSH. In Singapore, the MOM publishes the WSH National Statistics Report annually. Before interpreting the statistics, it is essential to know the changes in reporting requirements across the years. Firstly, as of 2014, the WSH statistics report:

a. included work-related traffic injuries;
b. reclassified work-related back injuries due to ergonomic risks, such as work-related musculoskeletal disorders, as occupational diseases;
c. expanded the number of workers to include those working in all workplaces covered under the WSH Act.

As such, the 2014 workplace injury rates are not strictly comparable with those from previous years.

In addition, from 1 September 2020 onwards, all work-related accidents and ill health need to be reported to the MOM. Before this, only accidents and ill health that resulted in more than four or more days of medical leave or more than 24 hours of hospitalisation needed to be reported.

The workplace fatal injury rate for all workplaces declined from 4.9 per 100,000 employed persons in 2004 to 1.2 in 2018 (Workplace Safety and Health Council and Ministry of Manpower, 2019). However, in 2022, the number of workplace fatalities rose to 46, and the fatal injury rate increased to 1.3 per 100,000 workers (Ministry of Manpower, 2023).

Fortunately, after a series of interventions by the industry and the Ministry of Manpower, the fatal injury rate reduced to 0.99 per 100,000 workers in 2023 (Kok, 2024). In 2022, about 80% of fatal and major injuries happened in the construction, manufacturing, transportation & storage, and some services industries. Despite improvements over the years, the construction industry has remained an area of concern. The persistently poor safety and health performance of the construction industry makes it especially important for all stakeholders to do their part to improve the situation.

Nevertheless, other industries, such as the process and energy industries, cannot be complacent; unlike the construction or marine industries, the process industry can cause major accidents with extremely severe consequences, ranging from hundreds or thousands of injuries and deaths. For example, the Bhopal gas leak disaster in India on 3 December 1984 killed at least 4,000 people. In addition, thousands of others were injured, and many others were ill. Major accidents like the Bhopal gas leak are stark reminders that major hazard industries must have high WSH standards.

1.3 Workplace Safety and Health Act

The Singapore Workplace Safety and Health Act (WSH Act) is heavily influenced by the UK Health and Safety at Work, etc., Act 1974, and similar legislations in Europe and Australia. As an illustration of the performance-based approach to WSH legislation, the WSH Act will be discussed herein.

The WSH Act was enacted in 2006 after three major accidents in 2004: the Nicoll Highway collapse on 20 April 2004, the collapse of steel latticework at the Fusionopolis Building worksite on 29 April 2004, and the fire on Almudaina at Keppel Shipyard on 29 May 2004. The overall intent of the WSH Act is to improve the industry's safety culture and encourage stakeholders to take reasonably practicable steps to improve WSH proactively. The WSHA is based on three basic principles:

1. Reducing risks at the source by requiring all stakeholders to eliminate or minimise the risks they create;
2. Instilling greater ownership of safety and health outcomes within the industries;
3. Preventing accidents through higher penalties for compromises in safety management.

Under the WSH Act, a workplace is essentially any place where a person carries out work. Some workplaces are classified as a factory, and they are usually high-risk workplaces. Factories are defined as premises where some articles or part of an article is being produced. The production process includes alteration, repair and decoration. The fourth schedule of the WSH Act gives examples of factories; they include manufacturing plants, car servicing workshops, petrochemical plants, shipyards, and construction work sites.

The predecessor of the WSH Act is the Factories Act. It is a prescriptive legislation that details WSH requirements specifically. In contrast, the WSH Act is a performance-based legislation that requires the industry to conduct risk assessment (RA) to proactively identify the hazards, evaluate the hazards, determine controls for the hazards, and implement and review the hazards and controls. One key feature of a performance-based approach is the concept of "as low as reasonably practicable" (ALARP). Based on the case of Public Prosecutor v Hong Jun Development Pte Ltd [2017] SGMC 68, "reasonably practicable" contains the following principles:

1. The risk of accident has to be weighed against the measures necessary to eliminate the risk, including the cost involved;

2. "The term 'reasonably practicable' means that stakeholders need only take preventive measures which are proportionate to the potential impact of the hazards at the workplace.";
3. "…if a precaution is practicable, it must be taken unless in the whole circumstances that would be unreasonable. And as men's lives may be at stake, it should not lightly be held that to take a practicable precaution is unreasonable."

In essence, "reasonably practicable" means that the benefits of reducing risk must be weighed against the costs (e.g., time, money and resources) of implementing the controls. The weighing or evaluation of costs versus benefits should refer to regulations, ACoPs, industry standards, guidelines and norms to define what is reasonable and practicable.

Compared to the prescriptive approach in the repealed Factories Act, the performance-based regime is more sustainable because the government cannot keep creating new regulations to cover different hazards and controls. The emphasis on WSH management is essential because penalising a company only when it has accidents or ill health is reactive, and workers would already be infected, injured or killed. Promoting effective WSH management reduces the risk of incidents and pollution.

The duty holders covered in the WSH Act and their key duties are highlighted in Table 1.1. Each duty holder has a role in preventing accidents and ill-health and can be taken to task for failing to perform their duties.

Besides the WSH Act, a set of subsidiary legislations are written under the WSH Act, and to support the subsidiary legislations and the WSH Act, a wide range of standards were developed. Standards approved by the MOM are known as ACoP and have a higher standing in the courts than non-approved codes

Table 1.1   Duties of duty holders under WSHA. (Adapted from (Ministry of Manpower, 2016b)).

Duty holder	Definition	Key duties
Employer	Section 6(1) — "employer" means a person who, in the course of the person's trade, business, profession, or undertaking, employs any person to do any work under a contract of service.	Section 12 — An employer must protect the safety and health of his employees or workers working under his direction, as well as persons who may be affected by their work. The employer must: • Conduct risk assessments to identify hazards and implement effective risk control measures. • Make sure the work environment is safe. • Make sure adequate safety measures are taken for any machinery, equipment, plant, article, or process used at the workplace. • Develop and implement systems for dealing with emergencies. • Ensure workers are provided with sufficient instruction, training and supervision so that they can work safely.
Principal	Section 4(1) — "Principal" means a person who, in connection with any trade, business, profession, or undertaking carried on by him, engages any other person otherwise than under a contract of service — (a) to supply any labour for gain	Section 14 — A principal must ensure that the contractor he engaged: • Is able to perform the work they are engaged for. • Has made sure that any machinery, equipment, plant, article, or process that is used at work is safe.

	or reward, or (b) to do any work for gain or reward.	However, if the principal instructs the contractor or the workers on how the work is to be carried out, his duties will include the duties of an employer.
Occupier	Section 4(1) — "occupier", in relation to any premises or part of any premises, means: (a) in the case of a factory where a certificate of registration has to be obtained in relation to the premises pursuant to any regulations — the person who is, or is required to be, the holder of the certificate; (b) in the case of a factory where a notification has to be submitted in relation to the factory pursuant to any regulations — the person who is named in the notification or is required to submit a notification; (c) in the case of any other premises — the person who has charge, management or control of those premises either on his own	Section 11 — An occupier must ensure that the following are safe: • The workplace. • All pathways to and from the place of work. • Machinery, equipment, plants, articles, and substances. The occupier must ensure that the above does not pose a risk to anyone within his premises, even if the person is not his employee. Section 19 — The occupier may also be responsible for the common areas used by his employees and contractors. Common areas include the following: • Electric generators and motors. • Hoists and lifts, lifting gears (chain, slings, and lifting chains), lifting appliances (chain blocks and small hoisting equipment), and lifting machines (cranes). • Entrances and exits. • Machinery and plants.

(Continued)

Table 1.1 (*Continued*)

Duty holder	Definition	Key duties
	account or as an agent of another person, whether or not he is also the owner of those premises.	
Manufacturer or supplier of machinery and equipment or hazardous substances	Not defined in WSHA	Section 16 — A manufacturer or supplier must ensure that any machinery and equipment or hazardous substances (see WSH Act Fifth Schedule) he provides are safe. He must: • Provide information on health hazards and how to safely use the machinery, equipment, or hazardous substance. • Examine and test the machinery, equipment, or hazardous substance to ensure that it is safe for use. • Provide results of any examinations or tests of the machinery, equipment, or hazardous substances.
Installer or erector of machinery	Not defined in WSHA	Section 17 — The installer or erector of machinery (see Fifth Schedule of the WSH Act) must ensure that the machinery and equipment that he has erected, installed or modified is safe and without safety or health risks when properly used.

| Employee | Section 6(1) — "employee" means any person employed by an employer to do any work under a contract *of* service. (Note that "contract *for* service" refers to "an independent contractor, such as a self-employed person or vendor, is engaged for a fee to carry out an assignment or project.") (Ministry of Manpower, 2017) | Section 15 — An employee must:
• Follow the workplace safety and health system, safe work procedures, or safety rules implemented at the workplace.
• Not engage in any unsafe or negligent act that may endanger himself or others working around him.
• Use the personal protective equipment provided to him to ensure his safety while working. He must not tamper with or misuse the equipment. |
| Self-employed | Not defined in WSHA | Section 13 — A self-employed person is required to take measures to ensure the safety and health of anyone in the workplace who may be affected by his work. |

of practice. The list of ACoPs is published in the Government Gazette.

The WSH Act imposes a maximum penalty of S$500,000 for a corporate body with a first conviction. For a repeat offender, this is increased to S$1 million (Ministry of Manpower, 2016a). For individuals, the maximum penalty is S$200,000 for a first conviction and S$400,000 for a repeat offender. Individuals can also be imprisoned for a maximum of two years. The MOM can also impose composition fines instead of prosecuting. Each offence may be compounded to a sum not more than half the maximum fine prescribed for the offence or S$5,000, whichever is lower.

1.4 Overview of workplace safety and health management

Cost, quality, competition, profit, timeline, environmental pollution, and WSH are different challenges that managers must handle daily. On the surface, these challenges are unrelated and independent. However, the reality is that they are intertwined and interdependent (see Fig. 1.2).

Fig. 1.2   The inter-dependent nature of all management issues.

For example, a project manager may instruct workers to bypass safety and health procedures to push for project progress. A safety inspection may be cancelled, or the safety barricades may not be installed to save time. This leads to higher risk, and if accidents happen, the workplace will be subjected to investigations, leading to delays and loss of profit. Quality and WSH are related because when workers rush their work, they are more likely to make safety-related errors and can easily make mistakes, leading to quality issues. Poor work quality can also mean future rework, which means more time lost. However, WSH incidents do not occur whenever WSH procedures and measures are breached. This is because there is always an element of uncertainty or luck. Thus, managers tend to focus on more certain — typically, more urgent — issues they face. This frequently leads to the neglect of WSH management. To counteract this tendency, managers need to focus on the fundamentals of WSH management, which align with good management and planning principles.

The three basic principles of the WSH Act show the importance of WSH management — in particular, Principle 3, which implies that companies can be penalised even if there is no accident on site. Organisations are expected to manage safety proactively, and all stakeholders must be involved. The fundamental principles of the WSH Act are also aligned with the well-known accident pyramid or ratio (see Fig. 1.3), which was developed based on a set of accident data that F. E. Bird collected (Bird *et al.*, 2003). The pyramid shows that there are disproportionately more minor injuries, property losses, and near hits for each major injury. This implies that if management continuously learns from near hits, property losses, and minor injuries, it is possible to prevent major injuries. Nevertheless, after the BP Texas oil refinery explosion in 2015, many academics warned that the concept of an accident pyramid could also be misleading. Major accidents, e.g., the major explosion of a plant and the collapse of a large structure, can have very different

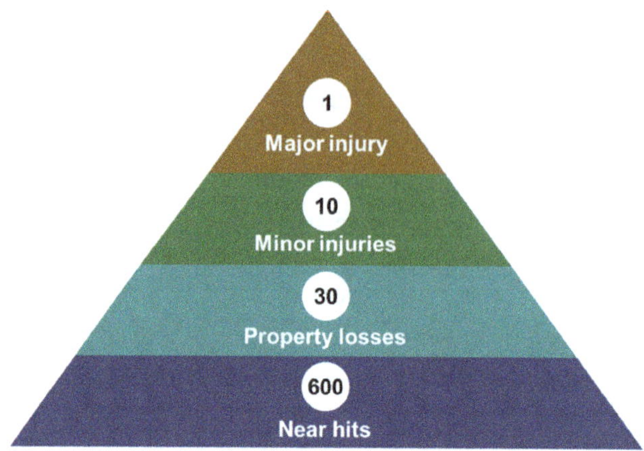

Fig. 1.3 F. E. Bird's accident pyramid.

causes from minor injuries, property losses, and near hits, especially if the nature of the incident is very individual, e.g., a worker cut his finger or fell when walking. The accident pyramid should only be taken as a guide. However, its emphasis on proactively identifying low-consequence incidents to learn from is still essential. In addition, the ratio of 1:10:30:600 is not fixed and can change when applied to different data sets.

WSH management is based on the Plan–Do–Check–Act (PDCA) cycle (Fig. 1.4), which is also adopted in quality, environmental, and risk management. During Plan, organisations must proactively identify the possible hazards that can lead to accidents and ill health. A series of controls or programmes will be identified during planning. During Do, the organisation will implement the controls and programmes to control the risk posed by the hazards and aspects. During Check, the organisation monitors the WSH controls and programmes using collected data. Organisations will also check that the plans are being implemented accordingly and evaluate if the processes are effective. If there are gaps, the organisation will Act on those identified issues for continuous improvement. The PDCA cycle continues to ensure progressive improvement.

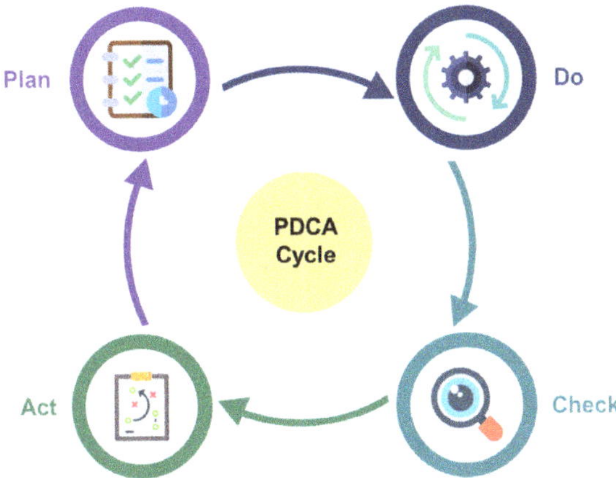

Fig. 1.4   The Plan–Do–Check–Act cycle.

In the context of WSH management, several standards and guidelines are available to help organisations understand the policies, structure, procedures, and actions needed to maintain a management system that continually improves WSH performance. Many WSH management system standards are structured based on the PDCA cycle. The ISO 45001:2018 Occupational Health and Safety Management Systems — Requirements with Guidance for Use (ISO 45001) (International Organization for Standardization, 2018) is an internationally adopted WSH management standard. ISO 45001 is similar to ISO 14001 Environmental Management Systems — Requirements with Guidance for Use (ISO 14001) (British Standards Institution, 2015), but the latter is designed for environmental management. ISO 45001 will be discussed in Chap. 6.

1.5 Workplace safety and health systems thinking

Systems thinking (or system dynamics) is a discipline that helps managers understand inter-relationships between system components to manage better complex systems that can behave unpredictably. One of the fundamental principles of systems thinking is the importance of looking beyond the direct and immediate causes of events. A systems thinker will seek to understand the behaviour patterns of the system, system components, and stakeholders to uncover the underlying factors and structure that caused the event. This is very much aligned with the concept of root cause analysis commonly found in WSH literature. However, systems thinking takes a broader view of underlying factors and structures than WSH root cause analysis. It goes beyond the WSH management system and safety culture to see the organisation as a whole system. Furthermore, systems thinking seeks to understand unpredictable or emergent system behaviour arising from interactions between system components. When necessary, the system's boundary being evaluated can also be expanded to cover areas beyond the organisation to facilitate the identification of underlying factors and leverage points.

This holistic view of WSH management is essential in improving WSH performance because WSH issues may have non-WSH origins. In addition, systems thinking emphasises that complexities arise from simple dynamics, such as the circular dependence between system parameters and components and delays (Goh *et al.*, 2010). However, we must note that system dynamics literature tends to emphasise quantitative approaches to model these complexities through stocks and flow models. In contrast, this book adopts the fundamental principles of systems thinking and takes a practical and qualitative approach to their application in WSH management.

Review questions

1. Why is WSH management important?
2. Why is workplace health more frequently neglected compared to workplace accidents?
3. How are WSH management and environmental management related?

4. Which are the industries that usually have poor WSH performance?
5. What are the three key principles of the WSH Act?
6. Explain the duties of the seven duty holders under the WSH Act.
7. Why is it essential for all stakeholders to emphasise the importance of WSH management?
8. What are the implications of F. E. Bird's accident pyramid?
9. Describe the PDCA cycle in the context of WSH management.
10. What is systems thinking, and how is it applicable to WSH management?

References

Bird, F. E., Germain, G. L., and Clark, M. D. (2003). *Practical loss control.* Georgia, Duluth: Det Norske Veritas (U.S.A.), Inc.

British Standards Institution. (2015). BS EN ISO 14001:2015 Environmental management systems — Requirements with guidance for use. London: British Standards Institution.

Goh, C. L. (2008). NZ engineer suspended from practice for two years. *The Straits Times.*

Goh, Y. M. and Soon, W. T. (2014). *Safety management lessons from major accident inquiries.* Pearson.

Goh, Y. M., Brown, H., and Spickett, J. (2010). Applying systems thinking concepts in the analysis of major incidents and safety culture. *Safety Science*, **48**, 302–309.

International Labor Organization. (2009). Facts on safety and health at work. Geneva: International Labor Organization.

International Labor Organization. (2023). Nearly 3 million people die of work-related accidents and diseases. https://www.ilo.org/resource/news/nearly-3-million-people-die-work-related-accidents-and-diseases

International Organization for Standardization. (2018). ISO 45001:2018 Occupational health and safety management systems.

Kok, Y. (2024, Jan 31). Workplace deaths in Singapore down 22% in 2023. *The Straits Times.* https://www.straitstimes.com/singapore/workplace-deaths-in-2023-fall-by-22 (accessed Jul 11, 2024).

Ministry of Manpower. (2016a). General penalties. 16 November 2017. http://www.mom.gov.sg/workplace-safety-and-health/workplace-safety-and-health-act/liabilities-and-penalties

Ministry of Manpower. (2016b). WSH Act: Responsibilities of stakeholders. 17 November 2017. http://www.mom.gov.sg/workplace-safety-and-health/workplace-safety-and-health-act/responsibilities-of-stakeholders

Ministry of Manpower. (2017). Contract of service. 14 November 2017. http://www.mom.gov.sg/employment-practices/contract-of-service

Ministry of Manpower. (2023). Workplace safety and health report 2022. Singapore: Ministry of Manpower.

Singapore Exchange. (2023). Sustainability reporting — Singapore Exchange (SGX). April 2023. https://www.sgx.com/sustainable-finance/sustainability-reporting

Workplace Safety and Health Council and Ministry of Manpower. (2019). WSH 2028. August 2019. https://www.mom.gov.sg/-/media/mom/documents/safety-health/publications/wsh2028-report.pdf

CHAPTER 2

Incident causation

2.1 Introduction

2.1.1 Definitions

Before discussing incident causation models, it is necessary to establish the difference between events, incidents, accidents, and near-hits (Fig. 2.1). An "event" is an episode of something that happened, which may or may not have consequences. Examples of an event include an overseas trip, a meeting, and watching a movie. "Accidents" are defined as unexpected and undesirable events that result in negative consequences such as injuries, fatalities, illnesses, property damage, and financial loss. Accidents are a subset of incidents, which include near misses or, more accurately, near hits — undesirable events that could have resulted in negative consequences had they not been averted by chance or successful safety or emergency measures. Based on the above definitions, "incidents" will include workplace accidents, environmental pollution, and occupational diseases.

2.1.2 Mental models

When an event occurs, different people may view the same event differently and can derive different insights from the same event. The differences in understanding can then lead to different actions with varying levels of effectiveness. These differences are influenced by the different mental models or

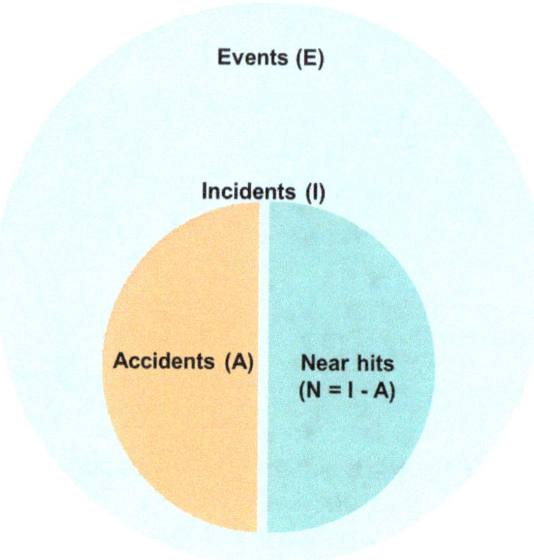

Fig. 2.1 Defining events, incidents, accidents, and near hits.

paradigms, which are our internal representations of how the world works. Mental models are made up of simplified concepts about the real world and the relationships between "things". Mental models are like maps, simplifying the real world to help guide our actions. Thus, when we manage workplace safety and health (WSH), it is vital to have a suitable incident causation model that guides us in preventing and investigating accidents. A good incident causation model will help us derive appropriate insights and actions to prevent incidents.

There have been many incident causation models developed over the years. This chapter will present the Domino Theory and Energy Transfer Model as traditional incident causation models, introduce the Loss Causation Model (LCM) and Swiss Cheese Model (SCM) as systemic incident causation models, and discuss the importance of systems thinking in incident causation.

2.2 Domino Theory

The Domino Theory by Heinrich (1936) (Fig. 2.2) is one of the earliest incident causation models, but it is now considered unsuitable. It focuses on how five dominoes are lined up from left to right. The theory posits that the collapse of the domino on the left, "Social Environment & Inherited Behaviour", leads to the failure of the next domino, "Fault of the Person", followed by "Unsafe Act or Condition", and finally leads to the collapse of the "Injury or Fatality" domino on the right.

Focusing on "Unsafe Act or Condition", an act refers to humans doing or not doing something. An unsafe act can be a worker doing something hazardous that can lead to an accident. For example, a worker removes a machine's guard or enters a dangerous area. Another common unsafe act is not wearing personal protective equipment (PPE) such as helmets, gloves, goggles, and safety boots. On the other hand, conditions refer to the state of the working environment, equipment, or material. For example, unsafe conditions can include slippery stairs and poor housekeeping, common in the construction industry. A faulty PPE, such as gloves with holes, would also be an unsafe condition.

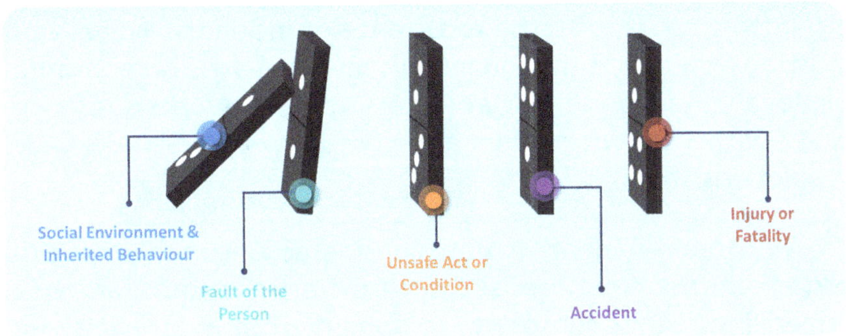

Fig. 2.2   An adapted version of Heinrich's Domino Theory.

According to the Domino Theory, an unsafe act is a critical domino in the chain of falling dominoes and is typically the fault of the person. According to the theory, if a person meets with an accident, it is because of his carelessness, recklessness, bad temper, etc., and this fault arises because of the person's upbringing, social environment, inherited behaviours, or even addictions like alcoholism. According to Heinrich, the fundamental cause of an accident, as represented by the last block of the domino, is the "Social Environment & Inherited Behaviour" of the victim and workers responsible for the accident.

The Domino Theory reflects the prevalent mental models of the past, which were usually focused on equipment, environmental factors, and human error (including violations). Such mental models persist today. For example, in July 2017, after the collapse of a viaduct under construction in Singapore, the media (Tan, 2017) highlighted human error as the likely cause of the accident without discussing the more fundamental management-related issues. Blaming the fault of persons, social environment, and inherited behaviour (non-management related factors) as the fundamental cause of incidents can lead to management blaming the workers for incidents and ending investigations before more fundamental or underlying factors are uncovered. The failure to remove or mitigate the underlying factors can lead to the recurrence of incidents. However, it must be noted that the individuals involved are also accountable for their actions and that it is essential for workers to cooperate with their employers and comply with WSH rules and regulations.

The Domino Theory is now considered unsuitable because organisations have realised that other more fundamental underlying factors contribute to human error, unsafe acts, and unsafe conditions. These underlying factors include poorly designed procedures, production pressure, error-provoking equipment design, and lack of management commitment to WSH. The general agreement among modern incident causation models is that even though unsafe equipment, environment and behaviours cause incidents, organisations are expected to manage these direct causes proactively and systemically.

2.3 Haddon's Energy Transfer Model

The Energy Transfer Model (ETM) (Haddon Jr., 1973) provides a practical understanding of incidents and control measures. The ETM (Fig. 2.3) traces the interactions between an energy source or hazardous substance and a person or property and the resulting consequences. Energy source refers to different types of energy like electrical, kinetic, mechanical, pressure, gravitational, thermal, etc. Hazardous substances include chemicals and substances that are toxic, corrosive, flammable, explosive, carcinogenic, radioactive, or biologically hazardous.

The energy source or hazardous substance has to be controlled with some form of control or barrier. When the person or property comes in contact with the energy source or

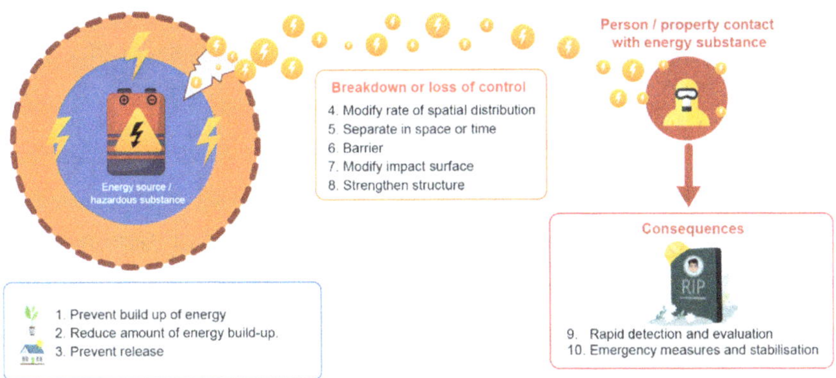

Fig. 2.3   The Energy Transfer Model. (Adapted from (Haddon Jr., 1973, Haddon's Energy Transfer Model)).

substance, there will be consequences such as injury or disease. Haddon's ten countermeasures or controls identified by Haddon Jr. (1973) are:

1. prevent the initial marshalling of the form of energy;
2. reduce the amount of energy marshalled;
3. prevent the release of the energy;
4. modify the rate of spatial distribution of release of energy from its source;
5. separate in space or time the energy being released from the susceptible structure;
6. separate the energy being released from the susceptible structure by interposition of a material barrier;
7. modify the contact surface, subsurface, or basic structure, which can be impacted;
8. strengthen the structure, which might be damaged by the energy transfer;
9. move rapidly in the detection and evaluation of damage and to counter its continuation and extension;
10. all those measures that fall between the emergency period following the damaging energy exchange and the final stabilisation of the process.

The abbreviated forms of the ten controls are in Fig. 2.3. The first two controls focus on the source of energy or substance. The first control aims to prevent a buildup of energy or an accumulation of hazardous substances. This is to eliminate the energy or substance. If elimination is impossible, you should control the build-up by reducing the amount of energy or substance accumulated. Using working at heights as an example, we prevent the build-up of energy by removing the need to work at heights. If that is not possible, we reduce the build-up of energy by restricting the maximum distance that the person can fall, using controls like a safety net and travel restraint.

The third control is to prevent the release of energy or substances. For example, if we put the chemical in a tank, the tank could be designed with a certain thickness or even be double-layered to prevent a release of the substance. If the energy or substance is being released, you try to control the rate of spatial distribution.

The fourth control is about the rate of spatial distribution. Spatial distribution is about how fast the energy or substance spreads. For example, if a tank containing diesel leaks, there should be a bund wall, which is a low wall surrounding the tank, to prevent further dissipation of the diesel. The wall forms a barrier to prevent the diesel from flowing into other areas.

The fifth control is to separate the person or property from the released energy or substance in terms of space or time. For example, if the person is near an open edge where the person can fall from heights, you ensure a certain distance between the person and the edge. The spatial distance could be enforced using a warning line or some barricades that prevent the person from reaching the hazardous edge, which can cause a sudden release of kinetic energy if the person falls over the edge. The other way is to separate the person from the energy or substance, using time; for example, a machine with moving parts can crush a person — sensors are installed so that the machine will stop if the person is in the vicinity. This means that the person's and the machine's activities are separated in time.

The sixth control is to separate the persons from the energy or substance using physical barriers such as fire-rated doors and walls. The seventh control is to modify the impact surface or structure of the released energy or substance. The focus is on the object transmitting energy, not the person receiving it; for example, padding possible contact areas of equipment or the airbags in vehicles — these controls reduce injury severity. The eighth control is to strengthen the structure of the vulnerable target. For example, you can strengthen a scaffold by having diagonal bracings and strengthen workers' core muscles to prevent back injuries through conditioning training.

The ninth control is rapid detection and evaluation, which allows prompt actions to reduce the consequences after the person contacts the energy or substance. Rapid detection and evaluation include prompt incident reporting procedures and communication systems, CCTV, and fire detection systems. Lastly, emergency measures and stabilisation include first aid, emergency medical procedures, and fire suppression systems to prevent the severity of the incident from escalating.

The ETM is helpful in risk assessment and incident investigation. Many incident causation models incorporate some of the concepts highlighted in the ETM. However, ETM is not a systemic model because it does not emphasise more fundamental and systemic causes of WSH incidents.

2.4 Systemic models

Systemic incident causation models focus on the underlying factors influencing the occurrence of an incident. These underlying factors are system characteristics, and they are usually more remote in terms of incident causation than unsafe acts and conditions. Different models define underlying factors differently, but most systemic models take a socio-technical view of underlying factors, and they typically refer to management systems (policies, procedures and structure) or control system weaknesses, leadership inadequacies, and organisational culture factors. Some models go beyond organisational boundaries and consider the influence of the industry, government, and societal factors (Rasmussen, 1997). However, this book focuses on preventing incidents at the organisational level, and factors beyond the organisation are not discussed.

Numerous systemic models have been developed over the years, for example, the Management Oversight Risk Tree (MORT) (Johnson, 1980), the contributing factors in accident causation (CFAC) model (Sanders and Shaw, 1988), the SCM (Reason, 1997), the LCM (Bird *et al.*, 2003), and the Modified Loss Causation Model (MLCM) (Chua and Goh, 2004). Nancy Leveson developed the STAMP (Systems Theoretic Accident Modelling and Processes) model (Leveson, 2004), which models accidents based on systems theory and focuses on three main components: constraints, hierarchical levels of control, and process models.

The Event Causation Technique (ECT), an extension of the MLCM, is the main systemic incident causation model that will be discussed and presented in detail in the following chapter. Systemic models reinforce the concept of multiple causation, where the cause of an incident does not lie in a single line of causation but often branches into various chains of factors. For example, when a worker touches a strong acid and gets burned, the accident is caused by multiple causes, e.g., the lack of safety gloves, safety knowledge, and supervision. Each of these causes can have deeper causes and factors.

Figure 2.4 captures the gist of systemic models. Essentially, events (including incidents) do not "just happen". They happen because of "patterns" across entities like people, machines and organisations and these patterns are observed across time. These could be seen as the "habits" or typical characteristics of people and organisations that were moulded by underlying factors like inadequacies in management systems, leadership, and culture. For example, an organisation that does not have a systematic procedure for conducting RA was not identified during management system audits (underlying factor). The lack of RA then led to hazards not being assessed and no control being implemented to reduce the risk of the hazards across all sites (pattern). This results in high-risk hazards, like open edges, being frequently observed on site, which leads to a worker

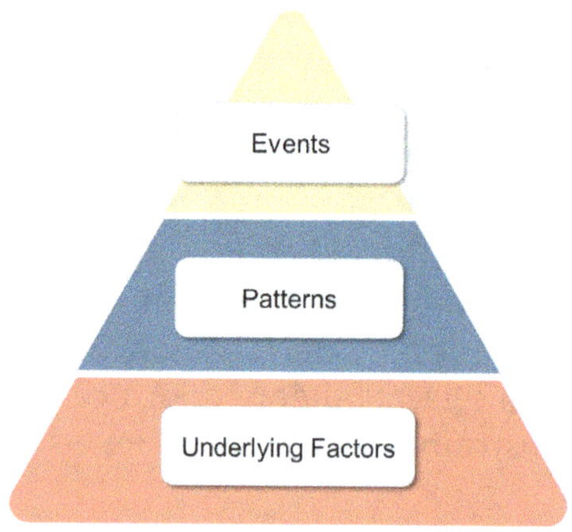

Fig. 2.4   Systemic view of events. (Adapted from (Senge, 2006)).

falling from height (an event) when working near one of the open edges.

2.5 Loss Causation Model

The LCM (Bird *et al.*, 2003) (Fig. 2.5) has a similar structure as the Domino Theory, but it represents a fundamental shift in the mental model of incident causation. Unlike the Domino Theory, the LCM deems "lack of management control" as the most fundamental cause of incidents and losses. Coming from

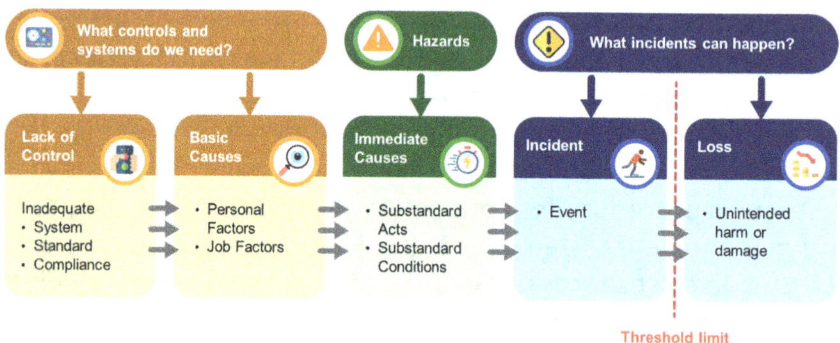

Fig. 2.5   Loss Causation Model. (Adapted from (Bird *et al.*, 2003)).

the model's left, lack of management control (inadequate management system, inadequate standard, or inadequate compliance to the system and standard) can lead to basic causes of incidents. As in the case of the Domino Theory, basic causes or root causes refer to personal factors and job or system factors. Personal factors are physiological and psychological, while job or system factors refer to factors such as leadership, work standards, RA, and maintenance. Table 2.1 presents different categories of each LCM component. Also note that the three arrows linking the different components indicate the concept of multiple causation, which was discussed earlier.

Immediate causes, including substandard conditions, directly cause incidents. "Acts" refer to observable human behaviours, and "conditions" refer to physical conditions such as the condition of equipment, material, structure, and environment. An act or condition is considered substandard when it fails to meet a stipulated standard set by the organisation. This implies that organisations are expected to set comprehensive standards to cover behaviours and workplace conditions. Immediate causes are more observable and more easily determined compared to basic causes.

"Loss" has a broader definition than human injuries and fatalities. Losses include property loss, reputation loss, and environmental impact. Following Haddon's ETM, losses arise from incidents when a certain protective threshold has been exceeded. Suppose a person is struck by a moving object, for example, a forklift travelling at a plodding speed. In that case, the collision may not exceed the person's capacity to absorb the kinetic energy exerted by the forklift. In such a situation, the incident would be a near-hit because the injury threshold of the person has not been exceeded, and there would not be any loss.

To better illustrate the LCM, Table 2.1 captures the possible classifications under each of the main blocks in the LCM. It must

Table 2.1 Loss causation model. (Adapted from (Bird *et al.*, 2003)).

Loss Causation Model component	Categories
Events / incident type	• Struck against (running or bumping into) • Struck by (hit by moving object) • Fall to lower level (either the body falls or the object falls and hits the body) • Fall on same level (slip and fall, tip over) • Caught in (pinch and nip points) • Caught on (snagged, hung) • Overstress/overexertion/overload • Caught between (crushed or amputated) • Contact with (harmful energy or substance) • Release of (harmful energy or substance)
Losses from incidents	• Injured worker time • Co-worker time • Leader time • General losses (lost production time, decreased effectiveness of employees, loss of business goodwill, etc.) • Property losses • Other losses
Substandard conditions	• Inadequate guards or barriers • Inadequate or improper protective equipment • Defective tools, equipment or materials • Congestion or restricted action • Inadequate warning systems • Fire or explosion hazards • Poor housekeeping; disorder • Hazardous environmental conditions: gases, dust, smoke, fumes, vapours • Noise exposure • Radiation exposure • Temperature extremes • Inadequate or excess illumination • Inadequate ventilation • Inadequate instructions/procedure

Table 2.1 (*Continued*)

Loss Causation Model component	Categories
Substandard acts or practices	• Operating equipment without authority • Failure to warn • Failure to secure • Operating at improper speed • Making safety devices inoperable • Using defective equipment • Using equipment improperly • Failing to use personal protective equipment properly • Improper loading • Improper placement • Improper lifting • Improper position for task • Servicing equipment in operation • Horseplay • Under the influence of alcohol and other drugs • Failure to follow procedure/policy/practice • Failure to identify hazard/risk • Failure to check/monitor • Failure to react/correct • Failure to communicate/coordinate
Lack of control (system elements)	• Leadership and administration • Leadership training • Planned inspections and maintenance • Critical task analysis • Incident investigation • Performance observation • Emergency preparedness • Rules and work permits • Incident analysis • Knowledge and skill training • Personal protective equipment • Health and hygiene control • System evaluation

(Continued)

Table 2.1 (*Continued*)

Loss Causation Model component	Categories
	• Engineering and change management • Personal communications • Team communications • General promotion • Hiring and placement • Materials and services management • Off-the-job safety
Job/system factors	• Inadequate leadership and supervision • Inadequate engineering • Inadequate purchasing • Inadequate maintenance • Inadequate tools and equipment • Inadequate work standards • Wear and tear • Abuse and misuse
Personal factors	• Inadequate physical/physiological capability • Inadequate mental/psychological capacity • Physical or physiological stress • Mental or psychological stress • Lack of knowledge • Lack of skill • Improper motivation

be noted that these classifications are not comprehensive and are written generically. They are helpful prompts when considering possible causes and factors during incident investigation and classification. However, incident-specific information must accompany the classification to provide the background information needed for someone to understand the reason for the classification.

2.6 Swiss Cheese Model

The SCM (Reason, 1997) in Fig. 2.6 is used in many industries, such as the oil and gas, aviation, mining, and nuclear industries. Based on the model, consequences are the result of several conditions: (1) the presence of a threat or danger that emits an accident trajectory, represented by the laser beam, (2) the presence of holes in the barriers created by active failures and latent conditions, and (3) the alignment of the holes to allow the accident trajectory to hit the "target", such as property, equipment or person. The barriers are synonymous with the controls in the LCM, and active failures are essentially the immediate causes. Latent conditions are like basic causes in the LCM. According to Reason (1997), active failures are the errors and violations committed at the "sharp end" of the system; that means by frontline workers, pilots, maintenance personnel, and operators. Active failures have relatively immediate and short-lived effects. Latent conditions are like pathogens to the human body, e.g., poor design, gaps in supervision, undetected manufacturing defects or maintenance failures, unworkable procedures, clumsy automation, shortfalls in training, and inadequate tools and equipment. These latent conditions can

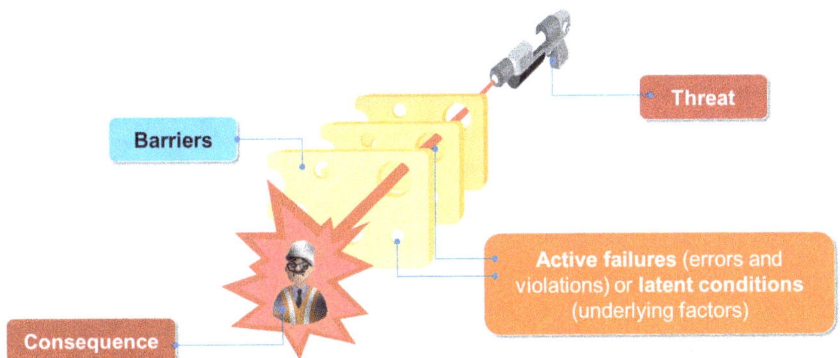

Fig. 2.6   Swiss Cheese Model. (Adapted from (Reason, 1997)).

be present in the organisations many years before they interact with active failures and local situations to defeat the different layers of barriers in the organisation. These latent conditions originate from high-level decisions of managers, designers, manufacturers, regulators, and governments.

Even though there are differences between the LCM and the SCM, both models emphasise that incidents and losses will occur if there is a lack of management control and an accumulation of underlying factors. However, the SCM highlighted safety culture as an important aspect of latent conditions, which the LCM did not cover explicitly. The concept of safety culture will be covered in Chap. 7. The SCM also emphasised that the holes in the barriers are inevitable, so the key is to reduce the number and size of holes as much as possible.

2.7 Systems thinking

Systems thinking seeks to understand how the different components of a system interact to promote the occurrence of events. When incidents happen, it is common for managers to focus on factors beyond their direct control or external to the system they manage, such as the poor WSH attitude of workers and supervisors, excessive market competition, and unsafe industry norms. Such a mindset is contrary to a systems thinking mindset. As Meadows and Wright (2008) discussed, many undesirable events like war, drug addiction, chronic disease, and environmental degradation are unintended, but they continue to happen. Similarly, WSH incidents are unintended, but they persist. These problems will only be eradicated when managers "stop casting blame, see the system as the source of its problems, and find the courage and wisdom to restructure it" (Meadows and Wright, 2008, p. 4). Therefore, systems thinking aligns with systemic incident causation models such as the LCM and SCM.

However, systemic incident causation models such as the LCM and SCM do not emphasise the importance of understanding the interactions between different parts of the organisation, such as management system, culture, and leadership. These interactions generate complex system behaviours and unintended events. This is where systems thinking helps generate insights and suggestions for dealing with different situations.

It is more evident that complexity can arise because of the unexpected interactions between physical components or production processes in high-risk systems such as nuclear plants or pharmaceutical plants. This is what Perrow (2011) focuses on when he highlights the importance of understanding how system accidents can happen because of the "unanticipated interaction of multiple failures" (p. 70). It is argued that WSH incidents are almost always complex; even the so-called straightforward incidents, e.g., falling from height and being struck by falling object, have significant complexities as one delves into the causes of the incident.

A system, in a more general sense, is a set of elements that interact with each other for a specific purpose. In the case of WSH management, the WSH management system includes policies, procedures and structures. The WSH management system also interacts with other management systems of the organisation, which have separate system elements. These management system elements interact with each other and the organisational culture (of which safety culture is a subset) and leadership to influence WSH-related behaviours and the occurrence or non-occurrence of WSH incidents. The interactions occur over time, and the cause and effect of the interactions and their subsequent influence on workers' behaviours are known as dynamic complexity (Senge, 2006), which are harder to comprehend, detect and mitigate. In contrast, detail complexity arises due to the large number of details, which are comparatively more manageable using a divide-and-conquer approach. Systemic incident causation models like LCM and SCM help identify the underlying management, cultural and leadership components that contributed to the incident (i.e., its detail complexity). However, systems thinking emphasises dynamic complexity that gives rise to an incident.

2.8 Causal loop diagrams

The discipline of systems thinking has a range of tools for uncovering the dynamic complexities of a system. A causal loop diagram (CLD) is one of such tools (Senge, 2006). Goh *et al.* (2010) introduced the use of CLDs in analysing incidents. In contrast to the linear nature of most incident causation models, a CLD can describe the circular nature of cause and effect and better explain the system's behaviour over time.

A CLD comprises three basic processes: a reinforcing feedback loop, balancing feedback, and delays. A reinforcing loop exists where a behaviour encourages similar behaviour in the future. It amplifies the behaviour over time, and an accelerating growth or decline will occur as the reinforcing loop continues. Figure 2.7 is an example of a reinforcing loop where the safe behaviour and its positive consequences, such as

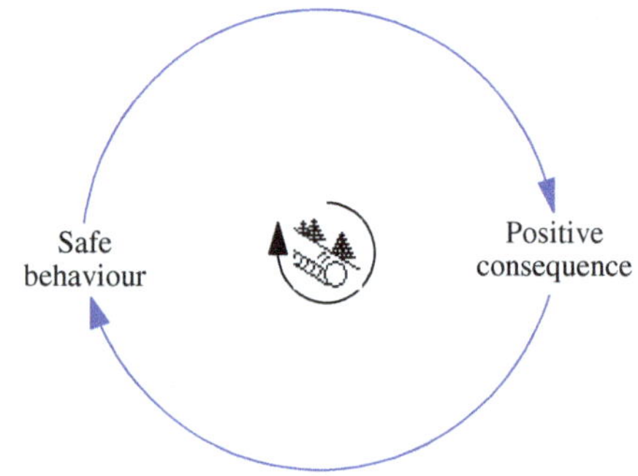

Fig. 2.7   Example of a reinforcing loop.

recognition by a supervisor and incentives from the company, create a virtuous cycle that encourages growth in the safe behaviour in the future, which then leads to further safe behaviour and positive consequences. A reinforcing loop can also be a vicious cycle where the unsafe behaviour (e.g., not doing safety checks) results in positive consequences like peer approval and compliments from "unenlightened" clients who want to avoid delays due to additional safety checks. These positive consequences will lead to more unsafe behaviour in the future.

Balancing feedback loops describe processes that aim to balance a behaviour or indicator at a target level. For example, as depicted in Fig. 2.8, a company may be monitoring a safety performance indicator ("Actual level"), and the company is motivated to increase its improvement effort whenever the target level is higher than the actual level. The larger the gap size, the greater the effort to improve the situation. The pressure to improve the safety performance reduces when the gap size reduces. To sustain the improvement effort, adjust the target level higher to maintain a healthy gap between the target and actual levels. The target level need not be explicit and can be implicit and hidden in people's minds. It is also possible for both explicit and implicit target levels to exist, and people may

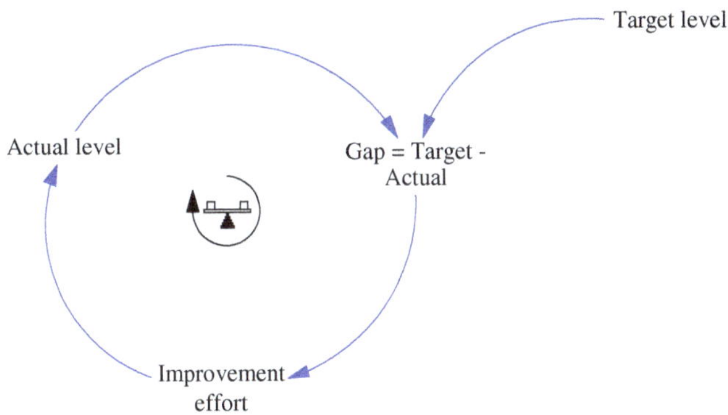

Fig. 2.8   Example of a balancing loop.

unknowingly be focused on the implicit target despite the explicit target level being higher.

The subsequent effect of a cause may be delayed. When a safety behaviour leads to a positive consequence, it might take some time for workers to be convinced that the safe behaviour can indeed lead to positive consequences (Fig. 2.9). Two parallel lines on the arrow linking two variables represent the delay. A lack of awareness of these delays can cause improvement efforts and safety programmes to be aborted prematurely.

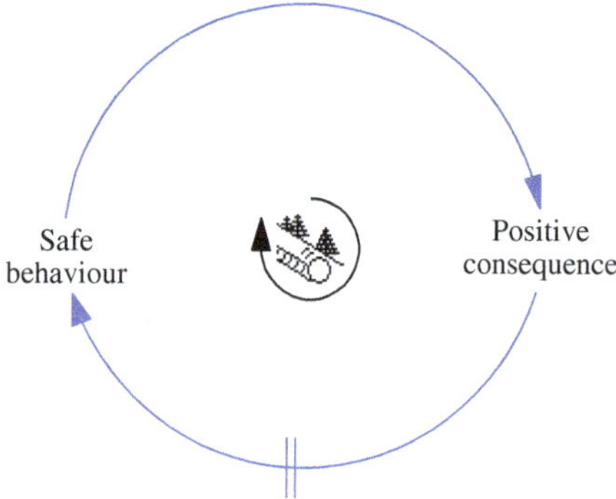

Fig. 2.9　Example of a delay in a reinforcing loop.

Using the basic building blocks of reinforcing loops, balancing loops, and delays, a series of systems archetypes can be captured to highlight common organisational problems and identify the possible leverage points organisations can focus on to improve organisational performance. The systems archetypes help understand the problems inhibiting WSH management and provide suggestions for improving WSH performance. These archetypes are described in (Senge, 2006) and (The System Thinker, 2018). The following paragraph will describe the "shift the burden" archetype (see Fig. 2.10) as an illustration of system archetypes. Other systems archetypes will be discussed in Chap. 3.

The archetype consists of two balancing loops, one addressing the symptoms and the other the underlying or fundamental

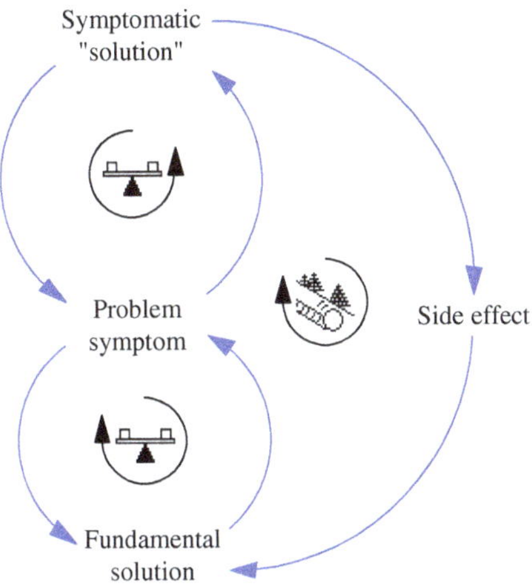

Fig. 2.10   Shifting the burden archetype.

causes of the problem. Symptomatic "solutions" can be very attractive to management, producing relatively quick positive results that focus on the symptoms and relieve the immediate pressure of the problem. The second balancing process focuses on fundamental solutions to the problem that are more sustainable but also have a delay between implementation and results. Failure to recognise this delay contributes to the tendency to focus on symptomatic solutions. When resources are focused on symptomatic solutions, the organisation's fundamental problem and capacity to resolve the problem worsens (side effect). Systems thinking concludes that short-term efforts aimed at the symptoms can be helpful, mainly to relieve the pressure as the delay between the fundamental solution and a manifestation of its benefits is experienced. However, long-term solutions to a problem must focus on the fundamental loop.

Archetypes help managers understand the possible underlying factors sustaining the pattern of system behaviour, which leads to the occurrence of the incident — the archetypes also present leverage points and possible strategies that managers can use to improve their current situation. The archetypes and

CLDs also allow managers to present their mental models to allow others to examine them and discuss solutions.

However, CLDs get too complicated very quickly. A well-known example is the CLD, developed to describe the complexity of American military strategy in Afghanistan (Bumiller, 2010). General Stanley A. McChrystal, the leader of American and NATO forces in Afghanistan, remarked, "When we understand that slide (with the CLD), we'll have won the war." From a communication point of view, CLDs are only helpful if the audience has been involved in developing the CLDs. It is a communication tool for a team to discuss and present members' mental models of the situation they are evaluating. Furthermore, developing a useful CLD takes significant skill and is arguably more of an art than a science. System dynamics simulation is another example of a systems thinking tool that cannot be easily used in practice. The simulation model quantifies all variables and requires statistical and mathematical knowledge that escapes most managers.

As highlighted in Chap. 1, this book will introduce the application of the philosophy and core concepts of systems thinking in WSH management, and more advanced tools, such as CLDs and system dynamics simulation, will not be discussed in detail. Even though systemic models do not surface dynamic complexity as well as CLD and system dynamics, they are easier to apply in real world context.

2.9 Conclusions

This chapter presented several incident causation models, and it was established that systemic models are now widely adopted by organisations. Systemic models emphasise that organisational factors like the management system, leadership, and safety culture are the fundamental causes of workplace incidents.

However, they do not emphasise the interactions between system elements, which give rise to dynamic complexity, making it more difficult for managers to identify and mitigate proactively. Thus, systems thinking was introduced as a helpful management approach that can be used to improve WSH management systems. The following chapters will describe different WSH management topics, highlighting relevant concepts of systems thinking to help readers understand how systems thinking concepts can be applied in WSH management.

Review questions

1. Describe the difference between an "event", an "incident", an "accident", and a "near hit".
2. Why is the Domino Theory no longer accepted in modern WSH management?
3. Apply Haddon's Energy Transfer Model to identify possible control measures to prevent a struck by lifted load accident in a construction site.
4. Give examples of events, patterns, and underlying factors of situations in your life. Demonstrate how the underlying factors cause the patterns and the events.
5. What is the critical difference between a systemic model (e.g., Swiss Cheese Model) and a non-systemic model (e.g., Energy Transfer Model)?
6. Explain the difference between immediate cause, basic cause, and lack of control in the Loss Causation Model.
7. Explain the similarities and differences between the Swiss Cheese Model and the Loss Causation Model.
8. Explain how detail complexity and dynamic complexity are different.
9. Explain how systems thinking can improve WSH management.
10. Search the internet for system archetypes using search terms such as "systems thinking archetypes". Explain how to select suitable archetypes for different system problems.

References

Bird, F. E., Germain, G. L., and Clark, M. D. (2003). *Practical loss control leadership*. Georgia, Duluth: Det Norske Veritas (U.S.A.), Inc.

Bumiller, E. (2010). We have met the enemy and he is PowerPoint. https://www.nytimes.com/2010/04/27/world/27powerpoint.html

Chua, D. K. H. and Goh, Y. M. (2004). Incident causation model for improving feedback of safety knowledge. *Journal of Construction Engineering and Management*, **130**(4), 542–551.

Goh, Y. M., Brown, H., and Spickett, J. (2010). Applying systems thinking concepts in the analysis of major incidents and safety culture. *Safety Science*, **48**, 302–309.

Haddon Jr., W. (1973). Energy damage and the ten countermeasure strategies. *Human Factors*, **15**(4), 355–366.

Heinrich, H. W. (1936). *Industrial accident prevention*. New York: McGraw Hill.

Johnson, W. G. (1980). *MORT safety assurance system*. New York: Marcel Dekker.

Leveson, N. (2004). A new accident model for engineering safer systems. *Safety Science*, **42**(4), 237–270. https://doi.org/10.1016/S0925-7535(03)00047-X

Meadows, D. H. and Wright, D. (2008). *Thinking in systems: A primer*. White River Junction, Vt: Chelsea Green Pub.

Perrow, C. (2011). *Normal accidents: Living with high risk technologies*. Princeton University Press.

Rasmussen, J. (1997). Risk management in a dynamic society: A modelling problem. *Safety Science*, **27**(2), 183–213. https://doi.org/10.1016/S0925-7535(97)00052-0

Reason, J. (1997). *Managing the risks of organizational accidents*. Ashgate, Aldershot.

Sanders, M. S. and Shaw, B. (1988). *Research to determine the contribution of system factors in the occurrence of underground injury accidents*. Pittsburgh: Bureau of Mines.

Senge, P. (2006). *The fifth discipline — The art & practice of the learning organisation*. Australia, New South Wales: Random House.

Tan, C. (2017). PIE uncompleted highway structure collapse: Cause likely to be human error, says veteran engineer. *The Straits Times*.

The Systems Thinker. (2018). Archetypes. https://thesys-temsthinker.com/topics/archetypes/

CHAPTER 3

Incident investigation

3.1 Introduction

Incident investigation is essential to any workplace safety and health (WSH) management system because it is important to determine the causes and underlying factors that led to the incident to prevent similar incidents. This chapter will cover the purpose of incident investigation, an overview of a typical investigation process, types of evidence, incident analysis, the Event Causation Technique (ECT), and how systems thinking concepts can be used to improve incident investigation.

3.2 ISO 45001:2018 requirements for investigation

According to ISO 45001:2018's Clause 10.2 Incident, Nonconformity and Corrective Action, organisations must "establish, implement and maintain a process(es), including reporting, investigating and taking action, to determine and manage incidents and nonconformities" (Chap. 6 provides an overview of ISO 45001). Based on the systemic model of incident causation (Chap. 2), incidents are not isolated events. We need to understand the patterns of the way things are done in the organisation as well as the patterns of behaviour across time, entities and people. These patterns help us understand the underlying factors that led to unsafe ways of working and, hence, the incident. There may be numerous underlying factors that contribute to the patterns and the events. Hence, an investigation should aim to:

1. Determine root causes (similar to basic causes and latent conditions discussed in Chap. 2) and contributory underlying factors.
2. Determine if similar incidents have occurred or could potentially occur.
3. Review existing risk assessments (RAs) and controls to assess their comprehensiveness and effectiveness.
4. Recommend corrective actions to prevent a recurrence.
5. Recommend continual improvement to WSH management, in general.
6. Communicate the investigation results to workers and other interested parties (also known as stakeholders).

Note that corrective action is defined as an action to eliminate the causes of a detected nonconformity (i.e., non-fulfilment of a requirement) incident or other undesirable situation and to prevent recurrence.

ISO 45001:2018 emphasises the importance of workers' participation and involvement of other relevant interested parties (e.g., contractors and suppliers) during an investigation. One of the key reasons is that the corrective actions developed will have to be implemented by the workers or the relevant interested parties, and it is critical for them to be involved in the development of corrective actions arising from the incident investigation. This will better ensure the feasibility and effectiveness of the corrective actions. Worker participation does not mean that workers must conduct the investigation themselves. Instead, they should participate through review and discussion of investigation findings. Other interested parties, e.g., clients, consultants and suppliers, may be interested in understanding how the organisation is dealing with the incident and seeking to improve safety management and prevent a recurrence. ISO 45001:2018 also requires organisations to react promptly to control, correct and deal with incidents' consequences.

During an investigation, it is common for the investigation team to identify opportunities for continual improvement that may not be directly related to the incident. These opportunities for improvement could be possible causes of the incident, which were proven unrelated to the incident. For example, when a team of investigators were investigating an incident involving a worker who tripped over a wire in a kitchen, they might realise that although the old tiles used in the kitchen did not contribute to the fall, non-slip tiles would help to reduce the risk of slips in the kitchen in the future. Hence, the investigation team needs to recommend non-slip tiles to ensure continual improvement of WSH management.

According to ISO 45001:2018, continual improvement is a "recurring activity to enhance performance". The recurring activity can be based on the Plan–Do–Check–Act (PDCA) cycle and does not have to arise from incident investigation. This means evaluating opportunities for continual improvement activities can occur at any time and need not be triggered by an incident. However, a thorough investigation process would typically be able to identify multiple opportunities for improvement. See Case Study 1 for examples of corrective action and opportunities for continual improvement.

Case Study 1 — Forklift Accident

A worker was struck by a forklift at a factory, leading to permanent injuries to the legs. The investigation found that the forklift driver was driving the forklift forward while carrying a load, which blocked the driver's view. The safe method was to reverse the forklift to ensure that there was a clear view of the path and any pedestrians in the vicinity. During the investigation, it was discovered that the worker who drove the forklift did not attend suitable forklift training and had driven the forklift without permission. Moreover, the forklift key was in the forklift and not placed in the key press as per procedure. A key press is a small

---

(*Continued*)

Case Study 1   (*Continued*)

cupboard meant to control access to keys through a logbook or key tracking system. There was no clear requirement on how frequently the supervisor, manager and workplace safety and health officer (WSHO) overseeing the key press should check on the keys in the key press.

Causes: Blocked vision when driving forklift ← (was caused by) unsafe driving ← untrained forklift driver ← uncontrolled access to forklift keys ← key not placed in key press ← inadequate monitoring and control of key press

Corrective actions for "inadequate monitoring and control of key press"

1. Purchase and implement a radio frequency identification (RFID) smart key press, which tracks and ensure all drivers return the keys to the key press whenever the forklifts are not in use. Supervisors, WSHOs and managers can monitor the keys based on the software that comes with the smart key press.

Corrective actions for "key not placed in the key press"

1. Make missing keys obvious with good visual markers in the key press.
2. The RFID system will send alerts to drivers, supervisors, WSHOs, and managers when keys are not returned on time.
3. All forklift drivers are briefed on the importance of removing the keys from the forklift and returning the keys to the key press. Disciplinary action will be taken if drivers are found to flout the safety rule. Incentives will be provided for drivers who comply with the procedure consistently.

Corrective actions for "untrained forklift driver"

1. All workers are reminded that unauthorised use of the forklift is dangerous and will be punished.
2. Authorised forklift drivers will be identified with a brown helmet with a name on it. Pictures of authorised forklift drivers are displayed on the notice board.

Case Study 1   (*Continued*)

> **Opportunities for continual improvement arising from this accident**
>
> 1. Preventive maintenance of the forklift will be implemented.
> 2. Additional pedestrian pathways will be created to reduce the likelihood of collision.
>
> The effectiveness of the measures will be reviewed by the safety committee 1, 3 and 6 months after implementation. Subsequently, a review will be conducted every 6 months as part of the biannual audit.

Corrective actions and opportunities for continual improvement should be prioritised based on the hierarchy of control (Chap. 4) and implemented through a management of change process (Chap. 6). Essentially, the hierarchy of control ranks controls based on their effectiveness in eliminating or reducing risk. Management of change refers to a structured process for the implementation of changes. Careful management of change for implementing new controls and improvements prevent new hazards from being introduced inadvertently. After the recommended actions are implemented, the organisation should review the effectiveness of the actions taken. This is necessary because not all recommended actions are effective, and there could be problems identified during implementation. The organisation should review the implemented controls and improve them.

The key purpose of an investigation is to prevent a recurrence and improve the management system and culture by eliminating or mitigating the systematic patterns of behaviour and management that produced the incident. As discussed in Chap. 1, incidents can be very costly. Thus, it is important that investigations go deep enough to identify patterns and underlying factors and not stay on the event level. This means that instead of simply blaming or even firing the workers involved in an incident for violating the safety rules, reactive event level responses, the investigation team should consider how the unsafe behaviour was facilitated by the situation that the workers were in. Understanding the underlying factors of unsafe behaviours facilitates proactive systems-level response. That said, it must be noted that disciplinary actions are still necessary to deter unsafe behaviours, but investigations should not focus on unsafe acts and punishments only.

3.3 Incident investigation and risk assessment

Figure 3.1 describes the role of incident investigation in relation to RA and WSH management system. Incident investigation is an indispensable reactive component of a WSH management system. It is reactive because an investigation is triggered only when there is an incident, which is undesirable due to its potential consequences. Nevertheless, incident investigations serve an important role in improving the way operations are being

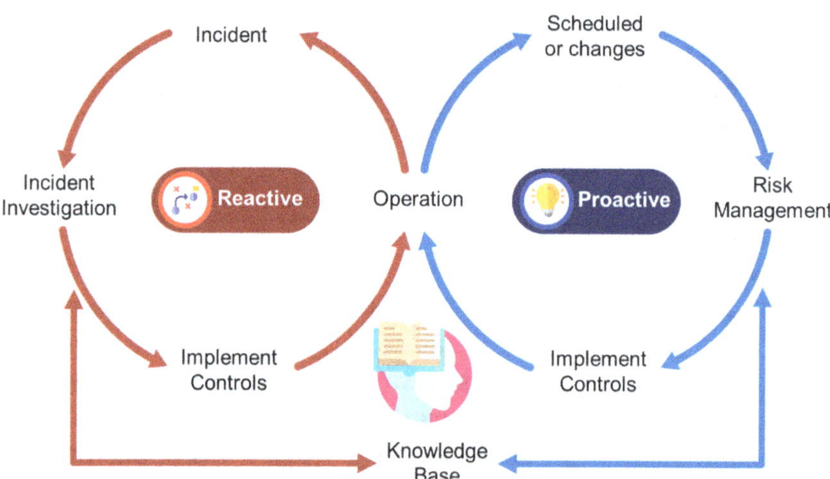

Fig. 3.1 Dependence between reactive and proactive control loops. (Adapted from (Chua and Goh, 2004)).

managed and controlled. On the other hand, the ideal situation would be for hazards and risk controls to be identified proactively so that operations are safe and incident-free. The key underlying process that promotes proactive WSH management is RA (Chap. 4). RAs are conducted prior to operations, at scheduled periods, after an incident, when there are changes in the operations (e.g., method, equipment and material), and when there is new WSH information that justifies a review. As depicted in Fig. 3.1, both proactive and reactive loops aim to improve the WSH performance of the operation.

To ensure that the organisation learns from its past experiences, it is important to capture the information generated during incident investigation and RA in a knowledge base so that investigators and RA (or risk management (RM)) teams can access the knowledge base when they are conducting investigations and RAs respectively. For example, it is important to consider the hazards and risk controls identified in the RA during an investigation. The investigators should assess if the hazards and risk controls were comprehensive and if the controls were implemented and effective. An incident investigation should make full use of the undesired incident to improve the underlying factors influencing the patterns of how operations are being conducted in the organisation. If not, the organisation would have missed an opportunity to learn. An organisation with a strong learning culture should have effective reactive and proactive control loops.

3.4 Hindsight bias

Dekker (2002) highlighted that investigators are prone to biases, especially hindsight bias. As represented by the person juggling the balls in Fig. 3.2, the people directly involved in an incident are always juggling different priorities. For example, a worker could be handling his work, which consists of different tasks and considerations that he must balance and keep up

Fig. 3.2 Investigators' potential hindsight bias.

with. He may also be thinking about issues in his personal life. In addition, as represented by the tunnel in Fig. 3.2, he does not have the full information of his work context, does not know what is at the end of the tunnel, and is not able to forecast the unfolding event. This implies that as the person walks inside the tunnel, looking forward, there is uncertain risk and incomplete information.

In contrast, investigators are on the outside looking backwards with clearer hindsight. This means investigators can see a more complete picture of the situation with greater certainty. This can lead to hindsight bias, where investigators fail to empathise with the challenges faced by the workers, which led to the mistakes. Hence, investigators should aim to understand the errors' rationale, consciously avoid blaming individuals, and assess the situation fairly. Frequently, the rationale for the errors and mistakes of individuals are the underlying factors of management system, leadership, and organisational culture.

As an illustration, in Case Study 1, the unauthorised forklift operator who took the key without authorisation could be balancing different work pressures while trying to make the best use of the time and resources that he was given. Even though it is still necessary to provide deterrence (i.e., punishment) for workers who commit such violations, it is more effective to prevent unauthorised access to the keys and reduce the likelihood of such violations. In contrast, managers can blame the operator for violating the safety rules, punish him, and assume that other workers will not violate the safety rules again after witnessing a worker being punished. Most of the time, such punishment will result in fear and can cause workers to stop providing WSH-related information to management. However, this does not mean that a "no-blame" policy will work. There is a need for fair punishment, which takes into account both the actual challenges that individuals face, and focuses on developing systems that prevent mistakes and violations. The need for a fair or "just culture" will be discussed in Chap. 7.

To avoid hindsight bias, investigators must be conscious of the bias and be critical of their findings. In addition, investigators should use incident analysis tools to help them structure their investigations to reduce the likelihood of biases.

3.5 Investigation team

Organisations should determine the different incident investigation roles and procedures before an incident happens. During an incident investigation, a manager should be appointed to oversee the investigation. The manager can be the manager in charge of the area where the incident occurred. However, if there are concerns about conflicts of interest, a manager from a sister unit with experience in similar operations should be appointed. Furthermore, a team of investigators familiar with operations and safety requirements should be appointed to conduct the investigation and make recommendations. Ideally, the investigator should be a supervisor, manager, or someone familiar with the work processes involved in the incident. An important investigation team member is the WSH professional, who acts as an advisor for the investigation. The WSH professional should advise on the choice of investigation techniques, and legal requirements. However, many organisations appoint WSH professionals to conduct investigations alone. This is unfavourable because the operations personnel should understand the work situation and incident better than WSH professionals, especially if the operations involve technicalities beyond the competency of WSH professionals.

Depending on the incident's complexity, the investigation team can be large. For example, for a major accident like a highway collapse or a vessel fire explosion, the team needs to organise the investigation like a project, with an overall incident investigation manager and personnel assigned to investigate specific aspects of the incident. Such large investigation teams are rare, and most investigation teams would only have one to three people. Regardless of the incident, it should be noted that the operations personnel should make up the main team, with WSH professionals engaged to support the investigation. This is only possible if the operations personnel are trained in incident investigation and have the right WSH mindset.

3.6 Investigation process

Investigation is an iterative process of collecting evidence, deriving the facts based on the evidence, generating opinions, and making judgements based on available facts. Typically, opinions and judgements will lead to a decision to collect other evidence. Thus, the loop continues until there are sufficient findings to determine the causes, underlying factors, and corrective actions. An investigation is also an information-filtering process. We select the information we think is relevant and focus on the selected information and continue to filter until we are confident that the findings are useful for preventing a recurrence. In some investigations, the focus could be on legal liabilities and fault finding; these are not the focus of this book, but the information presented herein are still useful for establishing accountabilities and liabilities.

The immediate priority after an incident is always the emergency procedures to mitigate the incident's consequences. Investigation will take place after the emergency response phase is over. Nevertheless, whenever possible, efforts should be taken to preserve the incident scene (e.g., restricting access to the incident scene and ensuring that objects related to the incident are left untouched) to facilitate the subsequent investigation process. It is important to conduct an RA prior to the emergency response to identify possible hazards to rescuers and provide necessary controls to assure their safety and health during the rescue. For example, prior to the rescue, it is important to shut down all machinery, isolate electrical power sources, review chemical safety data sheets, and provide suitable personal protective equipment.

An investigation can be split into four interrelated phases: (1) notify, (2) investigate, (3) analyse, and (4) report. Organisations need to be notified of incidents before an investigation can be conducted. During notification, the organisation needs to prioritise and allocate resources to the investigation. Some considerations when prioritising and allocating resources include the severity of the incident, the likelihood of a recurrence, the worst credible (or potential) severity of the incident, and the resources available. Many organisations launch investigations only when the incident severity is high, e.g., when a worker is hospitalised or severely injured. However, the actual severity of an incident is frequently a matter of luck. A worker struck by a falling wrench could have easily survived if the wrench fell one second later, and that would have reduced the incident consequence to zero. Thus, organisations should assess the worst credible severity of an incident and investigate if it is high. It is also important for the organisation to communicate with interested parties, e.g., government agencies, clients, workers and unions, on the investigation's purpose and process to allay concerns and garner support. A typical incident notification form is shown in Fig. 3.3.

Incident Notification Form

Date of Notification:	Time of Notification:
Date of Incident:	Time of Incident:
Location of Incident:	Type of Incident (injury/property loss/ill health/near hit/others):
Severity of Injury/Ill health:	
Potential Severity of Injury/Ill health (worst credible):	Activity Prior to Incident:
Likelihood of Recurrence of Incident:	High-Potential Incident? (Yes/No)
Type of Personnel Injured/Diseased:	Name and ID of Personnel Injured/Diseased:

Incident Description (attach photos and sketches):

Actions Taken:

Subsequent Actions:

Submitted by:

Fig. 3.3   Example of an incident notification form.

The investigation phase is a fact-gathering process intertwined with the "analyse" phase, which includes hypothesising about what happened, the causes and underlying factors, and evaluating the hypotheses against the facts gathered. The analysis process can involve the use of different methods and tools to guide investigators to probe into possible causes. Missing facts and evidence will be identified during the analysis and hence guide the investigation. As more facts and information are gathered, the hypotheses about the incident and its causes and underlying factors will be confirmed, amended or eliminated. New hypotheses can also be developed during the analysis. Thus, investigation and analysis activities are interrelated.

Based on the author's experience, the competency and mindset of the investigators and the management's commitment towards uncovering underlying factors are more critical than the incident analysis methods. A sophisticated tool or method, e.g., AcciMap and Functional Resonance Accident Model (FRAM), is useless in the hands of people who are not competent or not committed. Asking a series of "whys" is perhaps the most basic form of analysis, and it can be extremely effective if the investigators know how to frame the questions and the management is determined to uncover the systemic issues influencing the incident. However, if the investigators are competent, an advanced incident analysis method can help to make the investigation more structured and comprehensive. This chapter will provide a review of some of the established incident analysis techniques.

Finally, the reporting phase is about presenting the findings and recommendations to management and other stakeholders. A report is usually written and submitted to the management for discussion and approval. Many organisations use standard templates for investigation reports. An example is shown in Fig. 3.4, which comes with a set of incident investigation

Incident investigation form

Incident details		
Name of person involved in the incident:		Date of incident:
Location of incident:		

Incident investigation team:

What task was being performed at the time of the incident?

What happened? (e.g. 'employee tripped over box' or 'forklift hit wall')

What factors contributed to the incident?

Environment:		Equipment/materials:	
☐ Noise	☐ Layout / design	☐ Wrong equipment for the job	☐ Equipment failure
☐ Lighting	☐ Dust / fume	☐ Inadequate maintenance	☐ Material / equipment too heavy / awkward
☐ Vibration	☐ Slip / trip hazard	☐ Inadequate guarding	☐ Inadequate training provided
☐ Damaged / unstable floor	☐ Other	☐ Other	

Work systems:		People:	
☐ Hazard not identified	☐ No / inadequate risk assessment conducted	☐ Procedure not followed / no procedure exists	☐ Drugs / alcohol
☐ No / inadequate safe work procedure	☐ No / inadequate controls implemented	☐ Fatigue	☐ Time / production pressures
☐ Hazard not reported	☐ Inadequate training / supervision	☐ Change of routine	☐ Distraction / personal issues / stress
☐ Other		☐ Lack of communication	☐ Other

Corrective actions:

Contributing factor (from above list)	What are we going to do to fix the problem?	Who	When	Completion date

Issue fixed?		
Name	Signature	Date
Person involved in incident:		
Manager:		

Fig. 3.4   Incident investigation form by WorkCover Queensland (n.d.).

Incident investigation process guide

1. Establish the facts of the incident, including:
- What happened?
- When and where did it happen?
- What task was being done?
- Who was involved?
- Were there any witnesses?

2. Gather all necessary background information, for example:
- maintenance records
- safe work procedures
- instructions manuals
- training records

3. Consider all the potential contributing factors:

- Environment: *Did environmental conditions (e.g., light, noise, floor surfaces) contribute to the incident?*

- Equipment/materials: *Did anything about the equipment, materials, tools, etc. (e.g., equipment failures, missing guards) contribute to the incident?*

- Work systems: *Was there something about the system that contributed (e.g., hazard not identified, known hazard not addressed)?*

- People: *Was there something the workers, supervisors or contractors did that contributed to the incident (e.g., poor communication, being tired or rushing to finish on time)?*

4. Determine the primary cause/s of the incident; that is, those which if they hadn't occurred, then the incident wouldn't have occurred. Ask yourself, *"Would the incident have happened if….?"*

5. Identify the root cause / system failures that underlie the primary cause(s) and contributing factors.

One simple technique for identifying the root cause is the "Five Whys". This technique involves asking yourself "Why did this happen?" and continuing to ask "Why" for each response until you reach a conclusion that does not generate another "why" and the underlying cause becomes apparent.

6. The final and most import step in any investigation is to take action to fix all the factors that contributed to the incident, starting with the primary cause(s) and working through each of the contributing and underlying causes.

Fig. 3.4   (*Continued*)

guidelines (Incident Investigation Form (WorkCover Queensland, n.d.)). As shown in Fig. 3.4, the incident investigation report provides basic information about the incident and the injured and the task being performed at the time of the incident. More importantly, the report must contain the causes and underlying or contributory factors and corrective actions. The corrective actions should be tracked to ensure that they are implemented. A more detailed report will be needed if the incident is more complex, e.g., involves multiple parties, multiple machinery, and/or a complicated series of events.

It is important to develop and maintain a set of systematic and comprehensive investigation policies, procedures and templates. This includes a policy statement demonstrating clear management emphasis on the need to prevent the recurrence of incidents through thorough incident investigations. The policy statement should also be aligned with a systemic incident causation model by focusing incident investigations on fundamental management issues related to the organisation's management system, culture and leadership. Investigation resources, procedures, and supporting forms and templates, such as notification forms, investigation guidelines, a list of investigators for different types of accidents, pre-identified competencies required for different investigations, and pre-allocated resources for investigators (e.g., an investigation kit, financial resources, and time) should also be pre-determined. It is useful to involve operations staff in investigations as they are the most familiar with the operations. Workplace safety and health professionals who are more familiar with the WSH legislations and investigation processes can assist the operational staff and advise them on the investigation's purposes and processes. These roles and responsibilities must be pre-determined as part of the investigation procedure. The investigation guidelines provided by Workplace Safety and Health Council (2013) and Health and Safety Executive (2004) are useful examples of national guidance for investigation procedures.

It should be noted that the investigation process is not an exact science. The investigation team will frequently need to use their judgement and opinions to assess the facts and evidence gathered. However, it must be clearly stated when something is an opinion or a judgement, and the team should not mislead readers of the report into thinking that these opinions or judgements are factual. In addition, it is important to remind investigators of hindsight bias, as discussed earlier. They need to look beyond the individuals to consider the underlying factors contributing to the mistakes and errors. Knowing the reasons for human errors is critical in preventing a recurrence because a different person placed in the same situation can easily make the same error.

3.7 Types of evidence

There are four main types of evidence: part, position, people, and paper. "Part" refers to evidence in the form of material, equipment, and parts of the environment, such as a forklift's worn-out tyres, a vehicle's ignition key, environmental dust level or heaps of dust, noise level, illumination level, scaffold bracing, and fall arrest lanyards. "Part" evidence is usually collected from the incident scene. Thus, one of the first actions of an investigation is to cordon off the incident scene to prevent "part" evidence from being removed or altered.

"Position" refers to the spatial relationship between "part" evidence and the people involved and the placement of the evidence at different points in time. Some examples include the position of the injured before, during and after the incident, the location of the chemical safety data sheet before and after the accident, the position of the truck's wheel after the explosion, and the crane's position after the collapse. The investigator must take pictures of the position evidence to help document the incident scene. Sometimes, people near the incident location may have also taken pictures of the incident scene, related equipment, and activities. These pictures could be useful in understanding position and "part" evidence.

"People" evidence refers to statements provided by people with information about the incident. Some examples include interviews with the injured, witnesses (people who saw the incident), people associated with the activity, and people with information relevant to the investigation. Besides direct eye-witnesses and people directly involved in the activity, investigators frequently need to interview machinery suppliers, experts or experienced workers who understand how work is supposed to be conducted, and managers who are accountable for the work area in general. Interviews with witnesses and people with direct information about the incident should be completed as early as possible because people's memories tend to distort with time. In addition, witnesses might become affected by other people's intentional or unintentional comments.

Investigators must also obtain "papers" and documents (including electronic and video footage) related to the incident. These documents include risk assessments, standard operating procedures or method statements, job orders, emails, close circuit TV footage, safety data sheets (SDS), safety committee meeting minutes, shift change log books, permits-to-work, and crane data logger records.

The investigation team will have to decide what evidence needs to be collected. This is usually based on the investigation procedure, investigators' experience, the incident hypotheses, and the incident analyses.

Investigators must recognise the inherent fragility of evidence, emphasising its susceptibility to damage, alteration or loss. The compromise or loss of evidence can significantly hinder the results of an investigation. Evidence is fragile for various reasons. It can be intentionally tampered with, as seen when accident scenes are manipulated or when witnesses' memories are influenced by external factors. There are instances where evidence is deliberately destroyed or altered by individuals fearing the repercussions of an investigation. Unintentional distortions also occur, such as when chemicals or other substances are dispersed by natural elements like wind. Transient evidence, like quickly evaporating liquids, poses unique challenges due to its fleeting nature. Additionally, human errors can lead to accidental destruction or misplacement of evidence. One of the most common fragilities, however, stems from the malleability of human memory. For instance, a witness, after hearing others recount an incident, might inadvertently align their memory with these accounts, overshadowing their original recollection. Such vulnerabilities can disrupt the chain of evidence.

To ensure a comprehensive and unbiased investigation, the principle of triangulation should be employed. This approach dictates that each identified cause should be substantiated by at least two distinct pieces of evidence. Furthermore, investigators should rigorously crosscheck and validate each piece of evidence, comparing them across various categories of evidence (the 4 "Ps") and diverse sources, including different stakeholders.

3.8 Incident analysis methods and techniques

There are many incident analysis methods and techniques available in the literature. The Energy Institute (2008) captured 28 incident analysis methods that are suitable for analysing underlying (human and organisational) factors, but many methods were excluded from the document (e.g., AcciMap, Systems-Theoretic Accident Model and Processes (STAMP), and Man–Technology–Organization (MTO) analysis (Lappalainen and Perttula, 2017)). Most established incident analysis methods are useful in making an incident investigation more thorough and structured, but as indicated earlier, their degree of usefulness really depends on how the investigator makes use of the method and whether the management is committed to uncovering systemic issues.

Table A2, in the guidance provided by the Energy Institute (2008), differentiated 28 incident investigation methods based on the following features:

1. Whether training is required
2. Whether a software is required
3. Whether the method is designed to retrospectively evaluate existing investigation reports
4. Whether the method is used in the petroleum industry (or the relevant industry)
5. Whether the method requires the development of a pictorial representation of the incident
6. Whether the method is dependent on other tools to complete the analysis
7. Whether the method provides standard corrective actions and solutions
8. Whether the method provides checklists or flow charts to guide the analysis

These are important considerations when an organisation is selecting an incident investigation method. It should be noted that standard correction actions and solutions are usually not very useful because they are usually too general or basic.

Even though there are differences between the incident analysis methods, they also have significant similarities. For example, most of the techniques are guided by systemic causation models such as the Loss Causation Model (LCM) and the Swiss Cheese Model (SCM), which emphasise the importance of uncovering underlying factors, systemic issues, or root causes. The SCM is especially influential, and several methods, such as bow-tie, the Human Factors Analysis and Classification System (HFACS), and Incident Cause Analysis Method (ICAM), are based on the model. Another similarity across the techniques is the concept of multi-causation, where each event or cause arises due to a combination of multiple events and causes. Several methods emphasise the evaluation of "barrier analysis" (see Chap. 2 for a discussion on the SCM), where the target (people or property that can be harmed), barrier (control measures that prevent the target from being harmed) and energy (harmful energy or substance that can harm the target) have to be identified. Furthermore, a common feature of the investigation methods is the use of a timeline or chronological sequence of events. Last, each method typically has a set of prompts that guides the investigator in identifying possible events, causes, and underlying factors.

The author recommends the use of incident analysis methods that give experienced investigators flexibility in conducting their investigation and provide less experienced investigators with enough guidance to conduct their initial cases. Software-based methods can incur additional costs, and if the investigators are not used to the software, it may impede the brainstorming of possible hypotheses about the incident. Whiteboards, post-its and sketches are usually more useful during brainstorming and discussions, after which information can be summarised in a typical presentation, spreadsheet, or word-processing software. There are also diagramming software that facilitate collaborative work. The key features of a useful incident analysis method are summarised below:

1. Flexible and adaptable to the preferences and needs of the investigators.
2. Requires investigators to develop a chronological sequence of events.
3. Emphasises the concept of multi-causation, where investigators are encouraged to identify a comprehensive set of causes.
4. Prompts investigators to evaluate control measures that were or should have been identified prior to the incident.

5. Reminds investigators to uncover systemic issues (underlying factors) which contributed to the incident.

This book introduces the Event Causation Technique (ECT), which was created based on the features described above.

3.9 Event Causation Technique

The ECT is depicted in Fig. 3.5, and Table 3.1 shows the corresponding taxonomy. The ECT is based on the author's Ph.D. dissertation, and the technique was refined across the years after applications in actual accident investigations and analyses. The technique is an extension of incident causation models like the Energy Transfer Model (ETM), the LCM, and the SCM, but unlike these models, the ECT is meant to be a structured, simple, and flexible incident analysis technique. Investigators and managers using the ECT should be able to determine a comprehensive set of causes and factors to recommend actions to prevent the recurrence of an incident. It is also important to note that the ECT is meant to provide feedback to the proactive RA process (see Fig. 3.1), where the investigators are expected to review the control measures identified in the RAs relevant to the incident.

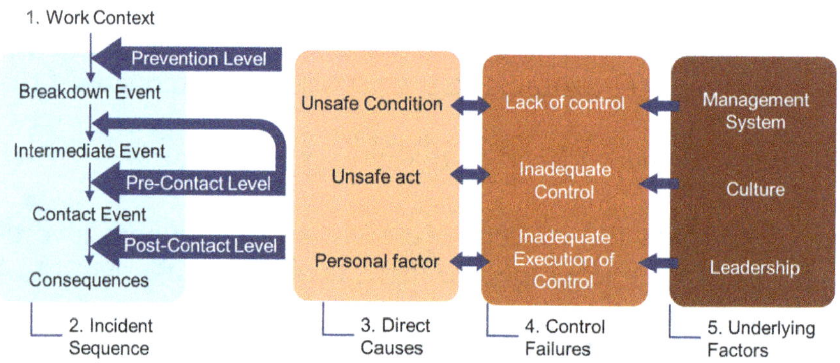

Fig. 3.5 Event Causation Technique.

Table 3.1   Event Causation Technique taxonomy.

Consequences		
1. Injured worker time	2. Co-worker time	3. Leader time
4. General losses (lost production time, decreased effectiveness of employees and loss of business goodwill, etc.)	5. Property losses	6. Other consequences
Incident events — Breakdown, intermediate and contact events		
1. Collapse of structure	2. Loss of balance	3. Release of (any harmful energy or substance)
4. Loss control of plant/equipment	5. Overstress/ overexertion/ overload	6. Fall to lower level (either the body falls, or the object falls and hits the body)
7. Fall on same level (slip and fall, trip over)	8. Struck by (hit by moving object)	9. Struck against (running or bumping into)
10. Contact with (any harmful energy or substance)	11. Caught in (pinch and nip points)	12. Caught on (snagged, hung)
13. Caught between (crushed or amputated)	14. Other events	
Direct causes		
Unsafe acts and practices		
1. Operating equipment without authority	2. Failure to warn	3. Failure to secure

(Continued)

Table 3.1 (*Continued*)

Direct causes		
Unsafe acts and practices		
4. Operating at improper speed	5. Making safety devices inoperable	6. Using defective equipment
7. Using equipment improperly	8. Failing to use personal protective equipment properly	9. Improper loading
10. Improper placement	11. Improper lifting	12. Improper position for task
13. Servicing equipment in operation	14. Horseplay	15. Under influence of alcohol and/or other drugs
16. Failure to follow procedure/policy/practice	17. Failure to identify hazard/risk	18. Failure to check/monitor
19. Failure to react/correct	20. Failure to communicate/coordinate	
Unsafe conditions		
1. Inadequate guards or barriers	2. Inadequate or improper protective equipment	3. Defective tools, equipment or materials
4. Congestion or restricted action	5. Inadequate warning systems	6. Fire or explosion hazards
7. Poor housekeeping; disorder	8. Hazardous environmental conditions: gases, dusts, smokes, fumes, vapours	9. Noise exposure
10. Radiation exposure	11. Temperature extremes	12. Inadequate or excess illumination

Table 3.1 (*Continued*)

Unsafe conditions		
13. Inadequate ventilation	14. Inadequate instructions/ procedures	15. Other unsafe condition(s)
Personal factors		
1. Inadequate physical/ physiological capability	2. Inadequate mental/ psychological capability	3. Physical or physiological stress
4. Mental or psychological stress	5. Lack of knowledge	6. Lack of skill
7. Improper motivation		
Control failures		
1. Failure to prevent the initial build-up of the energy or substance	2. Failure to reduce the amount of energy or substance build-up or stored	3. Failure to prevent the release of the energy or substance
4. Failure to reduce the rate of spatial distribution of release of energy or substance from its source	5. Failure to separate in space or time the energy or substance being released from the susceptible structure or person	6. Failure to separate the energy being released from the susceptible structure or person by interposition of an effective barrier
7. Failure to decrease the hazardous nature of contact surface, subsurface, or	8. Failure to strengthen the structure or person which might be	9. Failure to move rapidly in detection and evaluation of damage and to

(Continued)

Table 3.1 (*Continued*)

Control failures		
basic structure which can come into contact with susceptible structure or person	damaged by the energy transfer or contact with substance	counter its continuation and extension
10. Failure of all measures which fall between the emergency period following the damaging energy exchange and the final stabilisation of the process	11. Failure to ensure that relevant employees have sufficient awareness, knowledge, and/ or skills	12. Failure to provide adequate supervision
13. Other control failures		
Underlying factors		
Leadership (the buck stops here)		
1. Failure to take accountability	2. Failure to maintain oversight of WSH management system	3. Failure to maintain oversight of WSH culture
4. Failure to ensure adequate resources for WSH	5. Failure to communicate importance of WSH	6. Failure to protect workers who report incidents, hazards, risks, etc.
7. Lack of WSH knowledge	8. Lack of WSH leadership skills	9. Lack of commitment or inappropriate attitude or motiv-ation towards WSH

Table 3.1   (*Continued*)

Underlying factors		
Leadership (the buck stops here)		
10. Mental or psychological stress	11. Failure to demonstrate commitment to WSH	12. Other leadership factors
Management system inadequacies (based on ISO 45001:2018)		
1. Policy	2. Organisational roles, responsibilities and authorities	3. Consultation and participation of workers
4. Risk assessment	5. Determination of legal and other requirements	6. Planning processes
7. Resource allocation	8. Competence	9. Awareness and communication
10. Documentation	11. Risk controls	12. Management of change
13. Procurement	14. Emergency preparedness and response	15. Monitoring, measurement, analysis and performance evaluation
16. Internal audit	17. Management review	18. Incident, nonconformity and corrective action
19. Continual improvement	20. Other management system elements	

(*Continued*)

Table 3.1   (*Continued*)

WSH culture (Shared WSH values and beliefs in the organisation or at different organisational levels)		
1. WSH is given lower priority as compared to other organisational goals, e.g., productivity, deadlines and profit	2. Believes that incidents will not happen or insensitive to WSH hazards	3. Unsafe behavioural norms or practices
4. Other WSH culture issues		

The ECT has five main components: (1) work context, (2) incident sequence, (3) direct causes, (4) control failures, and (5) underlying factors. The following sub-sections will explain each of these components in more detail.

3.9.1 Work context

Work context describes the workplace scenario where the incident took place. Descriptors of work context can include the type of work being executed, the types of trades and workers involved, the equipment and material used in the work, the work environment where the work took place, and other interacting or nearby work. The descriptors can be used to categorise incident cases during incident data analysis. The work context, or parts of the work context, together with the direct causes, are the prerequisites of an incident.

3.9.2 Incident sequence

Incident sequence is a concept adapted from the ETM described in Chap. 2. It splits the incident into five types of key events:

preceding event, breakdown event, intermediate event, contact event, and consequence. The types of losses and events in the LCM are useful references for the incident sequence. The incident sequence describes what happened just before the incident till the occurrence of the consequences so that opportunities for preventing future occurrences of the incident can be identified.

A breakdown event is defined as an initiating point of loss of control of a source of energy or substance that, without an intervening event (e.g., presence of a control), will lead to the occurrence of a contact event. In contrast, a contact event is an event where the victim contacts the source of hazardous energy or substance. An intermediate event refers to any significant event between the breakdown event and contact event that, if adequately controlled, will prevent or mitigate the incident consequences. It is not compulsory to include an intermediate event (i.e., it is optional). The key criteria for deciding if intermediate events should be captured in the incident sequence is if there are opportunities for controlling the intermediate event. If the intermediate event represents an opportunity to put in controls to prevent future incident occurrences, then it must be described in the incident sequence. An intermediate event may also be included to ensure that the incident sequence is complete and logical to the reader, even if it does not provide an opportunity for control. In addition, there could be more than one intermediate event if the incident is complex. If additional intermediate events help the readers to understand the incident sequence better, then they should be inserted. That said, simplicity is always preferred, so intermediate events should only be inserted when necessary. At the end of the incident sequence are the incident's consequences. Consequences refer to the undesirable effects of the incident, e.g., property loss, environmental pollution, number of person-days lost, and type of injury. In most WSH investigations, the consequences would be focused on the injuries and ill health.

It is beneficial to define incidents based on the incident sequence structure so that causal factors and controls can be classified systematically into three levels of intervention and causation, namely, "post-contact", "pre-contact", and "prevention" levels. By focusing on the three levels of intervention and causation, investigators will be encouraged to broaden their scope of investigation to identify more opportunities for improvement. For example, when a worker loses their balance and falls off the edge of a building, an investigator could easily state that the "main cause" of the accident is that the worker was not using the fall arrest system provided. Even if the underlying factors and the controls that had contributed to the contact event (striking the ground) were identified, the recommendations based on the investigation would probably only prevent the recurrence of the contact event, but not the breakdown event (i.e., the worker's loss of balance, in this instance). To effectively lower the activity's risk, the factors that contributed to the occurrence of the breakdown event, the intermediate event, the contact event, and the consequences of the incident should all be identified.

The event just before the breakdown event, i.e., the preceding event, can be included to better describe the incident sequence. The preceding event can help to clarify the work context. It typically describes the actions of the workers involved in the incident before the breakdown event. The preceding event is unnecessary if the work context provides sufficient information about what happened before the breakdown event. It should also be noted that the preceding event should not overlap with the information captured under direct causes, e.g., an unsafe act by a worker.

3.9.3 Direct causes

As in the case of "immediate causes" in the LCM, direct causes refer to causes that are directly linked to an event in the incident sequence. A direct cause is a necessary condition for an event to occur. When the direct cause is removed, the event will not occur. Direct causes are classified into unsafe conditions, unsafe acts, and personal factors. Unsafe conditions refer to any direct cause related to non-human aspects, e.g., faulty equipment, unsuitable material, and unsafe work environment. Unsafe acts refer to observable actions or lack of action of front-line personnel executing the work. In most instances, the definitions of safe and unsafe acts are dependent on RAs and relevant standards and legislation. In broad terms, a condition or action that leads to an unacceptable risk level is unsafe. When standards and legislation are applicable, non-compliance is usually assumed to define what is unsafe.

Personal factors refer to factors such as the physiological, mental and psychological factors of a person (Bird *et al.*, 2003). Some examples include inadequate strength, insufficient physical endurance, lack of knowledge, and improper motivation. Personal factors that are direct causes of incident events are usually related to front-line workers or people directly involved in the incident. For example, consider a worker who performed an unsafe act by climbing up the bracings of an access scaffold and fell when climbing. A common personal factor that led to this unsafe act is improper motivation to save time and effort.

It should be noted that direct causes may not be present if the cause of the incident event is a control failure. This means that the incident event directly links to the control failure without a direct cause. This is because the control failure is the direct cause of the incident event.

3.9.4 Control failures

Control failures refer to failures of specific controls that an organisation has or omitted controls. These controls should have been highlighted in an RA that the organisation conducted prior to the incident. Failure occurs when:

- there is a lack of control, i.e., control was not identified in the relevant RA;
- there is inadequate control, i.e., control was identified, but it was not adequate in preventing the direct cause and/or incident event, or
- there is an inadequate execution of control, i.e., the control could have prevented the direct cause and/or incident event, but it was not implemented.

The possible control failures in Table 3.2, which were extracted from Table 3.1, are based on the ten countermeasures of the energy transfer model (Haddon Jr., 1973). The relevant incident events, i.e., breakdown event, intermediate event, contact event, and consequence, and corresponding level of intervention and causation were also identified in the table. Relevant ECT events refer to the incident event that can occur if the control failure occurs. However, with a wide variety of possible incident sequences and control failures, it is not possible for Table 3.2 to be comprehensive. Some examples of controls are edge barricades, warning signs, training, personal protective equipment, and emergency evacuation paths. The Hierarchy of Control will be explained in Chap. 4.

Controls are usually identified during an RA or RM process, which will be covered in Chap. 4. Controls need to be specific and must have a direct relation to the direct cause and/or incident event. A lack of control refers to the absence of control, for example, the absence of necessary barricades. Inadequate control refers to the presence of a control wherein the design or planning of the control was not adequate. For example, a barricade designed with insufficient height, where the open edge becomes inadequately protected, hence resulting in an

Table 3.2 Types of control failures.

BE — Breakdown event; IE — Intermediate event; CE — Contact event; CSQ — Consequence.

Possible control failures	Relevant ECT event (typical)	ECT level of intervention and causation	Relevant hierarchy of control
1. Failure to prevent the initial build-up of the energy or substance	BE	Prevention	Elimination
2. Failure to reduce the amount of energy or substance build-up or stored	BE, CSQ	Prevention, post-contact	Substitution
3. Failure to prevent the release of the energy or substance	BE, IE	Prevention	Engineering, administrative
4. Failure to reduce the rate of spatial distribution of release of energy or substance from its source	IE, CE	Pre-contact	Substitution, engineering
5. Failure to separate in space or time the energy or substance being released from the susceptible structure or person	IE, CE	Pre-contact	Engineering, administrative
6. Failure to separate the energy being released from the susceptible structure or person by interposition of an effective barrier	IE, CE	Pre-contact	Engineering

(Continued)

Table 3.2 (*Continued*)

Possible control failures	Relevant ECT event (typical)	ECT level of intervention and causation	Relevant hierarchy of control
7. Failure to decrease the hazardous nature of the contact surface, subsurface, or basic structure, which can come into contact with susceptible structure or person	IE, CE	Pre-contact	Substitution
8. Failure to strengthen the structure or perso, which might be damaged by the energy transfer or contact with substance	IE, CE, CSQ	Pre-contact, post-contact	Engineering, personal protective equipment
9. Failure to move rapidly in detection and evaluation of damage and to counter its continuation and extension	CE, CSQ	Post-contact	Administrative, emergency preparedness
10. Failure to implement all those measures that fall between the emergency period following the damaging energy exchange and the final stabilisation of the process	CSQ	Post-contact	Emergency preparedness

Table 3.2   (*Continued*)

Possible control failures	Relevant ECT event (typical)	ECT level of intervention and causation	Relevant hierarchy of control
11. Failure to ensure relevant employees have sufficient awareness, knowledge, and/or skills	BE, IE, CE, CSQ	Prevention, pre-contact, post-contact	Administrative
12. Failure to provide adequate supervision	BE, IE, CE, CSQ	Prevention, pre-contact, post-contact	Administrative

unsafe condition. Inadequate execution of control refers to non-compliance with plans, rules or designs related to a control. The unsafe act of using a non-explosion-proof torchlight in an area designated as a high fire or explosion risk zone can be due to an inadequately executed briefing (the control). In this example, the briefing could have failed to cover some of the required content, including the locations of the high fire-risk zones.

With reference to Fig. 3.1, it is important for investigators to evaluate the RA and the relevant controls. This will help improve the RA process's effectiveness, which is the cornerstone of any WSH management system. The information about control failures uncovered during an investigation will be useful inputs for RA teams selecting and evaluating controls. RA and RM will be discussed in Chap. 4.

Another aspect of control failures is the lack of Design for Safety (DfS) (Chap. 5) considerations when designing equipment, structures, machines, work environment, and other resources related to the incident. Investigators should consider how the design contributed to the control failure or how an inherently safer design could have prevented the accident. This is not commonly implemented in most incident investigations, but it could have a tremendous impact on improving WSH.

3.9.5 Underlying factors

Underlying factors refer to organisational and managerial factors that are contributory to the failure of controls and the occurrence of direct causes. They are similar to the latent conditions in the SCM and basic causes and lack of control in the LCM. Underlying factors are often contributory in nature, and their determination may have to depend on investigators' professional judgement, but identifying the underlying factors can lead to a broader and more sustainable impact on safety and health performance. Three types of underlying factors are identified in Fig. 3.5: leadership, management system, and culture.

In the context of WSH management, leadership refers to the attributes of senior managers that influence WSH. A senior manager is a person who has the ultimate authority and accountability, i.e., the buck stops with them. The leadership issues highlighted in Table 3.1 are possible factors contributing to the control failures and direct causes. Leadership issues will be discussed in more detail in Chap. 7.

A WSH management system refers to a set of policies, processes and structures that interact to assure or support WSH risk controls so that they maintain or improve the WSH performance of the organisation.

As an example, with reference to Table 3.1, a worker tripped over a wooden plank, i.e., the breakdown event is "loss of balance". The breakdown event was due to "poor housekeeping" (an unsafe condition), and the control failure is the "failure to conduct frequent housekeeping in the worksite" (i.e., failure to prevent the build-up of a hazardous "substance"). A relevant management system inadequacy could be inadequate processes for monitoring, measuring, analysing, and evaluating the system's performance. More specifically, although the site might have defined housekeeping schedules, there was a lack of inspections to check that housekeeping was actually conducted. At the management system level, the focus is on the pattern of events and not on the specific incident. Thus, it is important to do a wider check on whether the control failure is related to more widespread inadequacies in the relevant management system elements. Management system inadequacies are usually inadequacies in the PDCA cycle.

Another important underlying factor is organisational culture. Safety culture has gained a lot of attention in the WSH literature over the years. Safety culture can be seen as a subset of organisational culture, and the latter is defined as "a system of shared values [what is important] and beliefs [how things work] that interact with a company's people, organisational structures, and control systems to produce behavioural norms [the way we do things around here]" (Uttal, 1983). When applied to safety and health management, the "shared values", "[shared] beliefs", and "behavioural norms" refer to safety-related values, beliefs, and behavioural norms (see Table 3.1).

Organisational culture is abstract, and even people within the organisation may not be able to explain their own culture well, but they simply know that the culture is there. During the Safety Management in Context conference in 2013, Edgar Schein, an organisational culture guru, recommended that safety practitioners should focus on [management] processes and not "safety culture". He was sceptical about the concept of safety culture, which he felt was an ill-defined concept. If we take Schein's comments at face value, it seems to imply that the focus of any investigation of underlying factors should be on management processes. However, the author believes that Schein's comments were not meant to dilute the importance of organisational culture in incident prevention. His comments highlighted the importance of focusing on management systems so that shared values, beliefs and norms can be better managed.

It is true that the concept of safety culture is abstract, and its assessment is more subjective. Furthermore, culture constantly interacts with organisational structure processes, i.e., the organisation's management system. Therefore, a management system is a reflection of the culture and vice versa. Thus, from a WSH management perspective, it is more tangible to focus on the observable instead of focusing on the abstract. When applying the ECT, it is recommended that if there is sufficient information to infer the shared beliefs, values and norms that contribute to an incident, the cultural problems and issues should be highlighted. Safety culture will be discussed in more detail in Chap. 7.

3.9.6 Implementing the Event Causation Technique

To utilise the ECT, a series of steps should be conducted iteratively during the analysis stage of an investigation. The steps in the application of ECT are as follows:

1. Describe the work context in which the incident occurred.
2. Create a timeline of the incident. The timeline should be as comprehensive as possible.
3. Define the key events (i.e., breakdown event, intermediate event(s), contact event, and consequences).
4. For each of the key events in the incident sequence, ask a series of "whys" to facilitate the identification of direct causes, control failures, and underlying factors. In accordance with the principle of multi-causation, within each component of the ECT, there can be more than one answer to the "why". The asking of "why" questions should only stop when the answers reveal underlying factors, i.e., issues related to leadership, management system, and culture. If there are no direct evidence of the underlying factors, they can be inferred based on the other evidence collected.
5. For communication and presentation purposes, it is necessary to simplify the why-analysis into the ECT diagram (see Figs. 3.5 and 3.8) to ensure that readers can understand the incident causation in one diagram. This involves compressing the

why-analysis into more generic words and phrases (based on Table 3.1).

Table 3.1 contains useful prompts for the why-analysis and creation of the ECT diagram, where the description of causes and factors can be based on the categories in the table. At the same time, each cause and factor should be clear enough to help readers understand what the investigator is referring to. For example, if a relevant unsafe act is "failure to secure", the generic category should be followed by a brief description such as "did not secure dump truck tailgate".

3.9.7 Concrete hopper accident

The following case study demonstrates how the ECT can be applied in an incident investigation.

3.9.7.1 Describe the work context

In Case Study 2, a worker ("worker A") was working inside a man-cage attached to a concrete hopper. The concrete hopper is a container used to carry wet concrete. It has a release lever to allow workers to pour the concrete into formworks or other containers. In this case, the worker needed to pull the lever to pour the concrete into a joint between two precast wall panels. The concrete hopper was lifted by a tower crane, and the man cage was connected to the concrete hopper. The work context can be summarised into the following statement, "Worker pouring concrete into precast joint while standing in man-cage lifted by tower crane and using concrete hopper". The statement points out the type of activity being conducted and the equipment and plant that were used.

3.9.7.2 Create timeline

The timeline in Table 3.3 shows the events related to the case. The focus is typically on the events on the day of the accident,

Table 3.3   Timeline for Case Study 2.

Date	Time	Event	Remarks
1 Dec 2018	—	Project started	Project supposed to end in Dec 2021; contract between developer and contractor
13 Mar 2019	—	Worker A started working on site; induction training conducted.	Inducted by WSH Officer (WSHO); induction record and WSHO statement
1 May 2019	—	Hopper first installed with man-cage; crane operator started working on site	Inducted by WSH; see induction record and WSHO statement
May–Sep 2019	—	Man-cage used on several joint castings	Crane operator and lifting supervisor statements
2 Sep 2019	0830	Worker A starts work in man-cage; crane operator lifts man-cage	Co-workers and lifting supervisor statements
	0832	Hopper was lifted to the 6–7th level joint	Co-workers and lifting supervisor statements
	0835	Crane operator needs to adjust the height of the hopper to facilitate concreting	Co-workers and lifting supervisor statements
	0836	Hopper became dislodged suddenly	Co-workers and lifting supervisor statements
	0840	Site supervisor contacted WSHO	Site supervisor and WSHO statement
	0841	WSHO called ambulance	Site supervisor and WSHO statement
	0850	Worker A sent to hospital	Site supervisor and WSHO statement

but relevant information, such as the history of the hopper and man-cage, the start date of the project and other relevant activities, and the date when the operator and relevant workers first began working on site, can also be included. Furthermore, the relevant evidence and other remarks should also be captured in the timeline succinctly. The timeline serves as a useful summary of the key information for the incident, which is very useful for guiding the investigation and analysis phases.

3.9.7.3 Describe incident sequence

The incident sequence is created based on the timeline (Table 3.3). The events that are uncertain or were based on the opinions of investigators need to be highlighted using a symbol or a different colour. These uncertainties can then be further validated using new evidence, which becomes "to-dos" for the investigators. The incident sequence shown in Fig. 3.6 is the

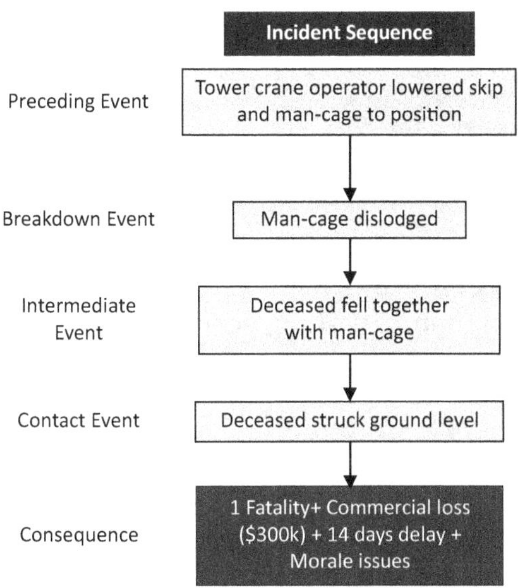

Fig. 3.6 Case Study 2 incident sequence.

finalised incident sequence. Investigators will face situations where there are several competing hypotheses about what happened during the incident, and several possible incident sequences can be drawn. Investigators need to use evidence to help them eliminate unlikely incident sequences and decide on the most likely incident sequence.

3.9.7.4 Conduct why-analysis

Once the incident sequence is established, the why-analysis should be conducted on each incident event. In Case Study 2, the discussion is focused on the breakdown event because during investigation, it was assessed that it was unlikely for the intermediate event and contact event to be prevented once the breakdown event is initiated. Furthermore, during the investigation, it was established that the emergency response was conducted promptly and in accordance with the emergency response plan. Sometimes, investigators can choose to focus on some of the key events due to time constraints, lack of resources, or unavailability of evidence, but it is emphasised that each incident event should be analysed as much as possible. With reference to Fig. 3.7, the qualitative analysis process is based on one simple question, "Why did the event or cause occur?" For instance, "Why did the man-cage dislodge from the concrete hopper?" In this case, the answer is "the man-cage dislodged because when the crane operator was lowering the hopper and man-cage into position (preceding event), the man-cage unhooked from the hopper (the cause for the breakdown event)".

Two basic guiding principles need to be consulted for each cause, the first being whether the cause is "necessary". The first guiding principle helps to check whether the cause or factor is a true cause. If it is not "necessary" (i.e., if it does not influence the existence of the event), then it should be removed. The second principle checks whether the cause is "sufficient".

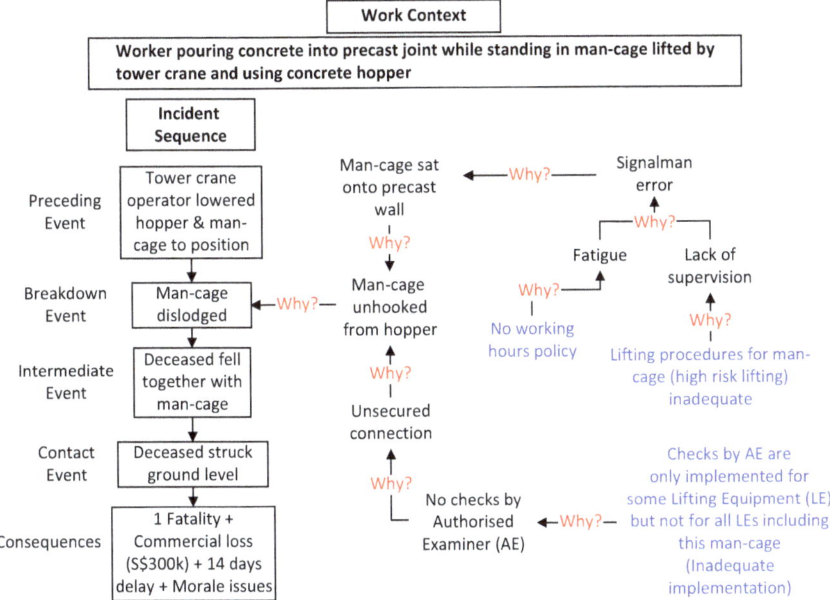

Fig. 3.7   Why-analysis for the breakdown event.

A cause may be necessary, but it may not be sufficient to cause the event, in which case another cause needs to be included. The sufficient condition supports the multi-causation concept discussed earlier.

In Case Study 2, the unhooking of the man-cage is necessary to cause the man-cage to be dislodged. This means that the man-cage will not be dislodged without the unhooking action. The unhooking is also sufficient to cause the dislodgement, i.e., other causes like strong wind or high temperature do not need to be present to result in the man-cage dislodging. The interrogation continues for "man-cage unhooked from hopper". In this case, two causes were identified. Firstly, during the lowering process, the man-cage sat on the precast wall, which produced a vertical reaction force, contributing to the man-cage becoming unhooked. Thus, "man-cage sat onto pre-cast wall" is a necessary cause. However, for the man-cage to "complete" the unhooking action, pushing the man-cage against the pre-cast wall alone is not sufficient to produce the "unhooking". The connection between the man-cage and the hopper must have been unsecured, which allowed a vertical force to push the man-cage out of the hopper. Thus, "unsecured connection" is included as another cause. The same process continues for each cause, guiding the investigator in his or her evidence-gathering process. The why-analysis can be guided by the taxonomy in Table 3.1, but the investigators should not be constrained by the taxonomy. The why-analysis should end when the investigator uncovers the underlying factors that can produce fundamental systemic improvement to the organisation.

Why-analyses are more effective if conducted in a team with a whiteboard or a wall with space for plenty of post-its. It is also possible to use a projector or a digital board with diagramming software to replace the physical post-its. The why-analysis should be reviewed as the investigation progresses and when new evidence is collected. If the analysis is done in a team, the facilitation skill of the investigator leading the why-analysis is critical to ensure its effectiveness. During the analysis, be it with a team or done individually, the investigator(s) must remember the following guidelines (GESIS) during the analysis:

1. Be **G**enerous initially and do not strike out any possible events, causes or factors.
2. Be **E**vidence-driven and highlight hypotheses or opinions not supported by evidence.
3. Be **S**pecific when responding to the "whys".
4. Be **I**nclusive and prompt different team members or other colleagues to provide inputs or critique the analysis.
5. Be **S**tructured as the team or investigator works through one event, cause or factor at a time and list actions or to-dos along the way.

3.9.7.5 Summarise using the Event Causation Technique diagram

The why-analysis can extend into several pages and can be confusing to the readers, especially personnel not directly involved in the analysis. Thus, the why-analysis is summarised into a simpler ECT diagram (Fig. 3.8). The ECT diagram should also include the summary of the analysis for the contact event and consequences. This is when each column in the ECT diagram, incident sequence, direct causes, control failures, and underlying factors, are filled in based on the why-analysis. Examples of ECT diagrams can be found in Chap. 10 and (Goh and Soon, 2014).

Using the ECT diagram, recommendations can then be made to improve the WSH performance of the organisation.

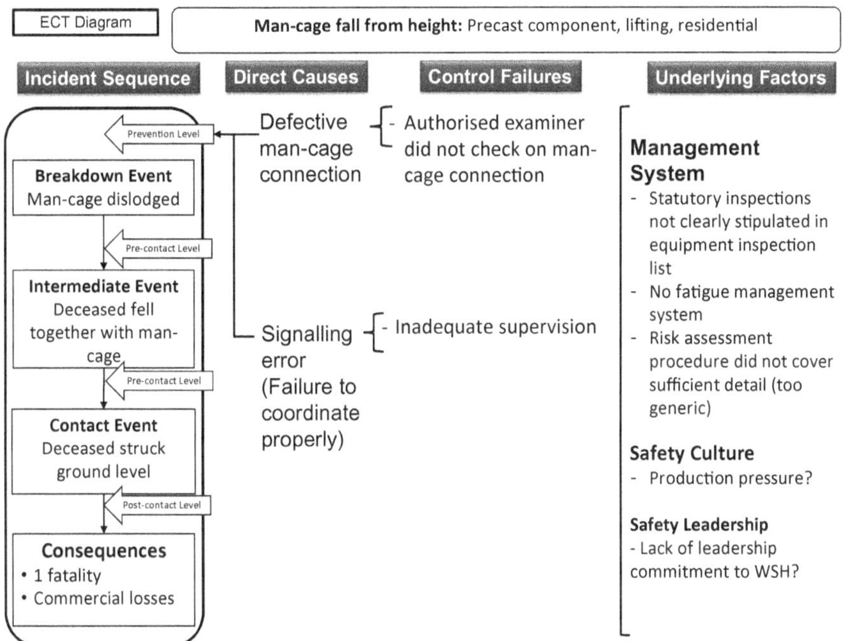

Fig. 3.8 ECT diagram for Case Study 2.

Each recommendation should be linked to different parts of the ECT diagram. Based on the ECT diagram for Case Study 2 (Fig. 3.8), some possible recommendations are:

1. Top management to impose an operational requirement for engineers, and supervisors to ensure that statutory equipment is checked before use.
2. Human resource department to develop a set of work hour policies and a fatigue management system that is applicable to all site personnel (including subcontractors); policies to be endorsed by top management.
3. To review lifting procedures and RA to differentiate between low- and high-risk lifting operations.
4. To review RA procedures to determine when generic RA is allowed.
5. To ban all man-cage usage until RA for lifting is reviewed; sites to use scaffold as a substitution in the interim.

The clear linkage provides a stronger rationale for the recommendation. The recommendations can then be monitored using an action plan that clearly answers the following:

- Who to do what and by when?
- When is the next review date?
- Who is the management representative to ensure progress?
- How to track effectiveness of measures?

3.10 Systems archetypes

Even when systemic models and investigation methods uncover underlying factors (leadership, management system, and culture) that contributed to the incident, the question of why these underlying factors arose still needs to be answered by management. As highlighted in Chap. 2, systems thinking literature (e.g., (Kim and Lannon, 1997)) has accumulated a set of systems archetypes, which are useful for identifying fundamental organisational dynamics that could lead to poor performance, including poor WSH performance. Chapter 2 highlighted one of the archetypes, "shifting the burden". Table 3.4 summarises ten archetypes and provides WSH examples and management guidelines related to the archetypes.

These archetypes are common organisational and management issues found in a wide range of contexts. By being aware of the systems thinking archetypes, management can identify possible leverage points they can focus on to improve WSH and other aspects of the organisation. Table 3.4 serves as a good primer for understanding deep-rooted dynamics contributing to workplace accidents and ill-health.

3.11 Summary

Incident investigation is an important part of WSH management. When an incident happens, there must be deliberate attempts to identify the patterns in the way things were done so that these patterns can be changed to prevent a recurrence of the incident. These patterns occur because of some underlying factors, such as the way the management system is implemented, the organisation's safety culture, and the leaders' safety commitment. Thinking of how the system influenced the occurrence of an event is a key concept in systems thinking and it is an important skill for any manager.

The investigation process is a qualitative, systematic and evidence-based process. The investigator needs to connect the investigation, a reactive process, with proactive RA, so that relevant risk controls identified in the RAs are reviewed, and future RAs strengthened. The ECT is a simple and flexible method for incident analyses. Despite its simplicity, it is meant to be thorough, and the final ECT diagram should provide a useful summary of an incident.

Table 3.4 Systems archetypes.

No.	Systems archetype	Description	WSH examples	Management guidelines
1	Balancing a process with delay	When working towards a goal, a party receives delayed feedback about the outcome of their effort. The party is not conscious of or does not have full knowledge of the delay. This leads to inappropriate actions (too much or too little) to close the gap between current performance and the goal. The party may also give up even before any significant results.	"We have not seen any improvement since implementation of the new WSH programme. Should we continue with it? Perhaps we should just cancel it."	• When feedback is delayed, be patient. • Build an information system that provides feedback quicker. • Actively seek the inputs of frontline staff.
2	Eroding or drifting goals	The two ways to close the gap between current performance and a goal are either to improve the performance or to reduce the goal. However, it takes	"It is fine to reduce the goal of having safety inspections everyday to once a month, as we need to get the work done by this week. We will remind	• Hold the vision, e.g., leaders need to reiterate the importance of WSH goals despite the difficulties on the ground.

		time and resources to improve performance, while it takes minimal effort to reduce the goal. This archetype describes how individuals or organisations take the easy way out of reducing the goal so that the pressure to improve is removed, but the actual performance is eroded over time.	the workers to be more careful." "Even though our goal is to review four risk assessments every month, we simply don't have the time. Let's review two this month and ask management to reduce the goal."	• Link goal setting mechanisms to parties not within the system so that it is harder to erode the goal, e.g., make WSH commitments and goals public and accountable to clients or unions.
3	Fixes that fail	There is a problem arising and it is usually very urgent. There is a quick solution to the problem, but it has negative consequences that will aggravate the problem.	"Our project team doesn't have the competency to conduct risk assessment. Let's employ consultants to conduct the risk assessment for us." "Everyone is afraid of the new virus and are clamouring for face masks. We should just give out masks to allay their fears. We'll deal with the shortage of face masks later."	• Use stopgap measures sparingly to "buy time". • Focus on long-term solutions. • Communicate the long-term consequences of fixes that fail.

(Continued)

Table 3.4   (*Continued*)

No.	Systems archetype	Description	WSH examples	Management guidelines
			"There are not enough technicians to implement the maintenance programme. Let's reduce the maintenance requirements so that it is quicker to finish each maintenance job. It might mean more breakdowns in the longer run, but let's deal with the workload issue first."	
4	Shifting the burden (addiction)	As an extension of fixes that fail and eroding goal — there are two options to solve a problem, either by using a fundamental solution that is more difficult to implement or use a short-term fix that has side effects	"It is too difficult to convince workers of the importance of WSH. Just punish those who flout the WSH rules even if it means that it will become harder to communicate with them in the future."	• Use stopgap measures sparingly to "buy time" • Focus on long-term or fundamental solutions • Communicate the long-term consequences of becoming "addicted" to the short-term solution

		on the capability to execute the fundamental solution. Organisations that get addicted to the short-term solution become unable to execute the fundamental solution in the longer term.		
5	Limits to growth	A party is experiencing growth, and the growth seems to be self-sustaining. At some point, the growth slows, a halt or even reverse into decay due to a "limit" or constraint in the system, e.g., resistance from different parties, resource constraint, lack of knowledge, and insufficient time.	"The behaviour-based safety programme was growing in terms of participation among the different project teams. However, the programme came to a sudden halt as some of the managers who were not involved perceived the programme as too expensive."	• Identify and focus on removing the limits early. • Do not push on the growth.
6	Growth but under-investment	A special case of limits to growth. Growth can be sustained if a party invests in its capacity to faciliate further growth. However, the investment to build	"Our customers are demanding us to do more in terms of WSH management, but that requires more investment in WSH training and	• Maintain continuous investment in WSH management capacity. • Be aware of the WSH management norms and stay ahead of the trends.

(Continued)

Table 3.4   (*Continued*)

No.	Systems archetype	Description	WSH examples	Management guidelines
		capacity is costly, and there is pressure to under-invest to conserve resources.	equipment. Let's wait and see if the demand from customers is just a passing fad."	
7	Escalation	When two parties are engaged in a "zero-sum" game where the gain of one party is perceived as the loss of the other party. In this situation, unhealthy competition can easily arise, and a vicious cycle can be created where the aggressive actions by one party triggers further aggression from the other party. If used wisely, the vicious cycle can be converted into a virtuous cycle, where the competition is used as an external motivation for improvement.	"The market is so competitive on price that everyone cuts their safety budget so much that safety performance is affected." "The other department is putting in a lot of effort into WSH; we should do the same; if not, we will look bad." "The other department is not doing anything on WSH and using all their resources to increase sales. We look like fools spending so much effort on WSH. We better do the same."	• Identify the indicator (e.g., price, sales figures, and WSH budget) that triggers the comparison and competition. • Change the indicator to promote a virtuous cycle or healthy competition (e.g., compete on a number of hazards reported with an acceptable level of quality). • Promote larger goals that emphasise collaboration (e.g., collaboration at the industry level to develop sustainable tendering practices).

		However, reliance on external motivation for improvement can expose an organisation to a possible vicious cycle that leads to poorer performance.	
8	Success to the successful	Two parties compete for limited resources, and the party that is more successful gets more resources, leading to more success. The other party gets fewer resources, leading to less success.	"Unlike the other manager, this manager is doing a great job in implementing the WSH measures. Let's give him a few more WSH coordinators to help him with his interventions."	• Bears some similarity with Escalation. • Promote common goals and teamwork so that the successes are shared. • Stipulate different goals for different teams so that the definition of success can be differentiated and de-coupled.
9	Accidental adversaries	Two parties cooperate to increase the chances of success but end up limiting each other's success. The degeneration starts when one party takes action that has unintended negative consequences on the other party, which leads to	"We are supposed to cooperate with company B on WSH initiatives so that our clients can see a consistent approach across the whole supply chain, but whenever we meet out clients, they always brag about their WSH initiatives,	• Ensure open communication between the partners. • Evaluate the possible impact on the partnership prior to any key actions.

(Continued)

Table 3.4   (*Continued*)

No.	Systems archetype	Description	WSH examples	Management guidelines
		retaliations and, hence, a vicious cycle.	which makes us look bad. We have to do something about this."	• Make the benefits of the collaboration explicit to each party.
10	Tragedy of all common goods (tragedy of the commons)	Two or more parties share a common resource that is limited. However, each party uses the resource from their own perspective without considering that the resource is limited. The common resource can end up depleted, and all parties are impacted.	"The centralised WSH team is being overloaded because every department is relying on them to conduct risk assessment, safety inspection, and other WSH management tasks. Now their support is so weak that it's like they are non-existent." "Everyone assumes that the world has unlimited supply of fish, clean air, water and other natural resources. Everyone keeps polluting the air, cutting down trees and digging up natural resources, and that is why we have climate change."	• Manage the shared resources by making the use of the resources explicit and rationing the resources. • Make each party self-regulate and become more independent of the shared resource. • Educate all parties about the limitations of shared resources.

Even when the underlying factors are identified, the management will need to understand the system dynamics that led to the underlying management system, leadership, or cultural problems. One way to help the management understand the system dynamics is to evaluate the ten systems thinking archetypes highlighted in this chapter. The systems archetypes describe common management issues and provide guidelines for resolving underlying factors and system issues.

Review questions

1. "An accident is a matter of luck." Discuss the validity of this statement and how management should see the role of luck in accidents.
2. Explain the difference between corrective actions and actions for continual improvement.
3. Explain how the incident investigation and RA are related.
4. Explain what it means for investigators to understand the "retrospective" nature of incident investigation.
5. Provide examples of the four types of evidence in an incident investigation.
6. Explain the differences among the breakdown event, intermediate event, contact event, and consequence.
7. Conduct a why-analysis on a problem you are facing.
8. Practise the ECT method on an accident case you find online.
9. Compare and contrast the different systems thinking archetypes. Which of them apply to your own situation?

References

Bird, F. E., Germain, G. L., and Clark, M. D. (2003). *Practical loss control leadership*. Georgia, Duluth: Det Norske Veritas (U.S.A.), Inc.
Chua, D. K. H. and Goh, Y. M. (2004). Incident causation model for improving feedback of safety knowledge. *Journal of Construction Engineering and Management*, **130**(4), 542–551.

Dekker, S. (2017). *The field guide to understanding "human error"*. Florida: CRC Press.

Energy Institute. (2008). Guidance on investigating and analysing human and organisational factors aspects of incidents and accidents. http://www.energyinstpubs.org.uk/tfiles/1368180505/817.pdf (accessed: 10 May 2013).

Goh, Y. M. and Soon, W. T. (2014). *Safety management lessons from major accident inquiries*. Singapore: Pearson.

Haddon Jr., W. (1973). Energy damage and the ten countermeasure strategies. *Human Factors*, **15**(4), 355–366.

Health and Safety Executive. (2004). *HSG245 Investigating accidents and incidents — A workbook for employers, unions, safety representatives and safety professionals*. London: Health and Safety Executive.

Kim, D. H. and Lannon, C. P. (1997). *Applying systems archetypes*. California: Pegasus Communications Inc.

Lappalainen, J. and Perttula, P. (2017). Accident investigation techniques. https://oshwiki.eu/wiki/Accident_investigation_techniques#cite_note-13 (accessed: 1 September 2020).

Uttal, B. (1983). The corporate culture vultures. *Fortune*, 17 October, 66–72.

WorkCover Queensland. (n.d.). Incident investigation form. https://www.worksafe.qld.gov.au/data/assets/word_doc/0019/131761/incident-investigation-form.doc

Workplace Safety and Health Council. (2013). *Workplace safety and health guidelines — Investigating workplace incidents for SMEs*. Singapore: WSHC.

CHAPTER 4

Workplace safety and health risk management

The Event Causation Technique

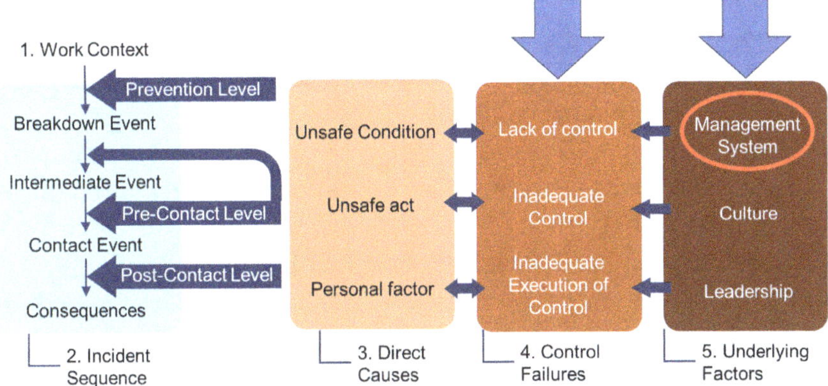

Risk management (RM) is an important aspect of any management system and it provides the core processes for a workplace safety and health (WSH) management system (see the Event Causation Technique (ECT) diagram above). One of the key deliverables of RM is the controls to prevent incidents and reduce the severity of injuries and ill health.

4.1 Introduction

Risk management (RM) is not unique to workplace safety and health (WSH) management. According to the ISO standard, BS ISO 31000:2018 Risk Management — Guidelines (British Standards Institution, 2018), risk is the effect of uncertainty on objectives. The following definitions and comments are adapted from BS ISO 31000:2018 and ISO Guide 73:2009 (British Standards Institution, 2009):

- An effect is a deviation from the expected, and it can be positive, negative, or both.
- The effect can address, create or result in opportunities and threats.
- Objectives can be about different aspects of an organisation, such as financial, productivity, WSH, and environment. At the same time, objectives can be at different levels of an organisation, e.g., the strategic, project, product, and process levels.
- Uncertainty is the partial or complete deficiency of information related to understanding or knowledge of an event, its consequences, and its likelihood.
- Risk is usually expressed in terms of risk sources (or hazards), potential events (occurrence or change of a particular set of circumstances), their consequences (in relation to the objectives), and their likelihood (or chance).
- RM is a systematic process of identifying, analysing, evaluating, controlling and monitoring risk. It is implemented to help ensure the achievement of organisational objectives.

In the context of this book, the objectives we are concerned with are WSH objectives. When organisations assess WSH risk, they typically refer to negative effects such as accidents and ill health, but there are also potential positive effects, like higher employee morale, higher productivity, improved reputation, and new business opportunities. However, these positive effects are usually only apparent at organisational or strategic levels. Most WSH RM is conducted at the operational level and focuses on operational issues, which are usually negative events.

We will first discuss the importance of RM in WSH and the role that it plays. We will then cover the RM process and key

WSH hazards and controls. Finally, we will discuss some of the problems and difficulties with WSH RM.

4.2 Workplace safety and health risk assessment and management

RM, which includes risk assessment (RA), is a cornerstone for a per-formance-based WSH legislative structure. The term RM is some-times used interchangeably with RA, but some differences will become obvious when we discuss the steps in RM subsequently.

4.2.1 Background

One of the key documents that resulted in the common use of risk assessment in WSH is the United Kingdom's (UK) Roben's report in 1972 (Committee on Safety and Health at Work, 1972), which highlighted the following:

- health, safety and welfare at work could not be ensured by an ever-expanding body of legal regulations enforced by an ever-increasing army of inspectors;
- the primary responsibility for ensuring health and safety should lie with those who create risks and those who work with them;
- the law should provide a statement of principles and defini-tions of duties of general application, with regulations set-ting more specific goals and standards.

Singapore adopted a similar approach with the enactment of the WSH Act in 2006. At the time of the enactment, then Minister of Manpower Dr. Ng Eng Hen highlighted the follow-ing principles of the WSH Act:

- Reduce risks at source
- Promote industry ownership of standards and outcomes
- Penalise poor management (ensure management accountability)

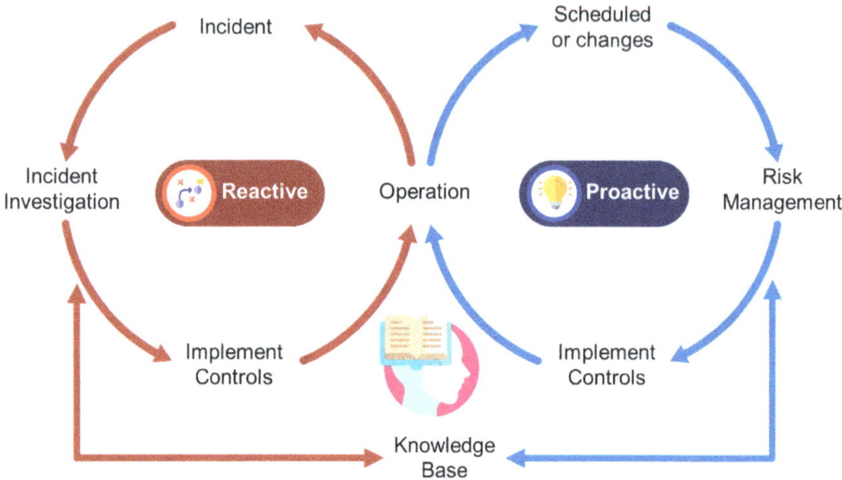

Fig. 4.1   Risk assessment allows stakeholders to be proactive.

Thus, for the industry to reduce risk at source and own standards and outcomes, employers must conduct RA (as part of their RM) to proactively reduce the WSH risk of its operations and activities (see Fig. 4.1). Although incident investigation plays a crucial role in WSH management by providing insights to prevent future occurrences, its reactive nature means it typically follows incidents, some of which may have led to significant injuries or losses. Thus, instead of incident investigation, RA should always be the key focus of WSH management. RAs must be scheduled before all operations so that controls can be implemented. RAs should also be conducted whenever there are changes in the operations, after significant incidents, and when organisations receive additional or new WSH information that sheds light on hazards and possible unintended events, and their likelihoods and potential consequences.

Furthermore, an incident investigation and RA generate WSH knowledge that must be properly stored in a knowledge base. The knowledge base can be used to provide inputs to future RAs and incident investigations. For example, RA teams should refer to past incidents when identifying hazards and assigning risk levels. Investigators must also refer to relevant RAs and assess the effectiveness and level of implementation of the controls identified in the RAs. Hence, both reactive and proactive learning loops are important, and organisations need to accumulate the knowledge gained from both loops into a knowledge base for sharing within the organisation. This knowledge base is usually a software that collects WSH data and provides dashboards, analytics and reports to help managers understand the WSH status of the organisation.

4.2.2 Risk assessment and incident causation models

The RA process is very much related to incident causation models. This is because incident causation models tell us why and how incidents happen, and RA aims to identify possible incidents, hazards and controls (see Fig. 4.2). During RA, it is especially important to identify the controls to manage the basic causes and immediate causes (hazards) to prevent incidents from happening.

The possible controls can be guided by Haddon's Energy Transfer Model (ETM) (see Chap. 2) and the hierarchy of control, which will be discussed later.

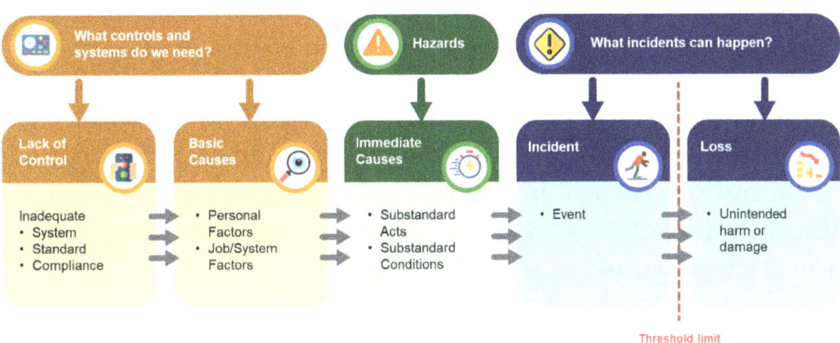

Fig. 4.2   Risk assessment and incident causation model.

4.2.3 Conceptual basis of risk assessment

The UK Health and Safety Executive (HSE) published the seminal document "Reducing Risk, Protecting People" in 2001. It provides a useful conceptual basis for WSH RA. Some key concepts include:

- Risk is a given. There are hazards everywhere, even if we do not do anything. Thus, it is theoretically impossible to have zero risk.
- Numerous factors influence risk perception of man-made hazards. For example:
  - if a hazard poses a risk to things that are important to us, we will perceive the hazard to be a higher risk;
  - if we do not understand a process giving rise to the hazard, i.e., we have less knowledge and competence, we will perceive the hazard to be a higher risk;
  - if the risk is not equitably distributed, we will perceive the hazard to be a higher risk ("Are we the only ones that are exposed to the risks?");
  - if we cannot control our exposure to the hazard, we will perceive the hazard to be a higher risk ("Do we have the resources and capabilities to control the hazards?");
  - if we assumed the risk involuntarily, we will perceive the hazard to be a higher risk ("Did we choose to be exposed to this risk or was it imposed on us?").

As seen, RA is heavily influenced by human minds, cultures, judgement, and values. Thus, it is difficult to establish the "objective and true risk", which may not even exist. Ultimately, it is about the tolerability and acceptability of risks by individuals, organisations and society. In addition, a major consideration is the resources (e.g., time, personnel, equipment, and money) that should be invested to control the hazards. The resources used in hazard control will have to be weighed against the aggregated incident consequences suffered by society, communities, organisations, and groups. Thus, RA is essentially a cost–benefit analysis, which balances the costs of controlling risk and the benefits of having the risk controlled. However, RA should not become an excuse to not implement risk controls. Organisations and individuals are expected to be proactive, conservative and comprehensive, such that more weight is given to WSH risks than the cost of controls.

As highlighted in Chap. 1, a key guiding principle of RA is "as low as reasonably practicable" (ALARP). The risk levels of all hazards should be ALARP, which means that the costs involved in reducing the risk level further would be grossly disproportionate to the benefits gained. Figure 4.3 describes how a hazard's risk level can be categorised into "Broadly acceptable", "Tolerable" and "Unacceptable" regions. The Unacceptable region is where the risk level of a hazard is too high, and no person should be exposed to this risk regardless of the benefits gained. On the other hand, the Broadly Acceptable region indicates that the risk level is widely regarded by society to be acceptable because the inherent risk is low or the controls are very effective. The Tolerable region is where the risk level is significant but tolerable such that the

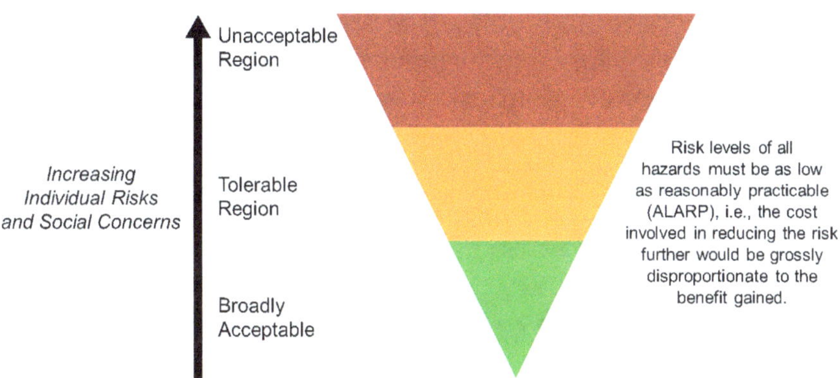

Fig. 4.3   HSE Framework for Tolerability of Risk. (Adapted from http://www.hse.gov.uk/risk/theory/r2p2.pdf, p. 42).

benefits justify the hazard exposure. Some benefits include employment, lower cost of production, personal convenience, and continued supply of basic needs like energy, food and water. In each of these regions, the risk level of the hazard should be kept ALARP. The concept of ALARP is especially important in the Tolerable region. For a hazard to be tolerated, it must be properly risk-assessed, the residual risk level must be ALARP, and the risk level must be periodically reviewed to ensure that it remains ALARP.

4.3 Risk management regulations

We will discuss the Singapore WSH (Risk Management) Regulations 2007 (RM Regulations) as an example of RM legislation or framework. In Singapore, all employers, self-employed and principals[1] who engage contractors are required to comply with the RM Regulations. The RM Regulations is a concise legislation containing only eight regulations, but it is a fundamental regulation in the WSH regime. Anyone who employs or engages others for work has to conduct RA (see RM Regulations). RA is defined as "the process of evaluating the probability and consequences of injury or illness arising from exposure to an identified hazard, and determining the appropriate measures for risk control". A "hazard" is defined as anything with the potential to cause bodily injury and includes any physical, chemical, biological, mechanical, electrical or ergonomic hazard. According to the RM Regulations, "risk" refers to the likelihood that a hazard will cause a specific bodily injury to any person. This definition is more specific than that provided by ISO 31000 because it is contextualised to WSH.

---

[1] Under the Workplace Safety and Health Act, "principal" means a person who, in connection with any trade, business, profession, or undertaking carried on by him, engages any other person otherwise than under a contract of service (i.e., an employee), to supply any labour for gain or reward or to do any work for gain or reward. Typically, a principal refers to the client in a client–contractor relationship.

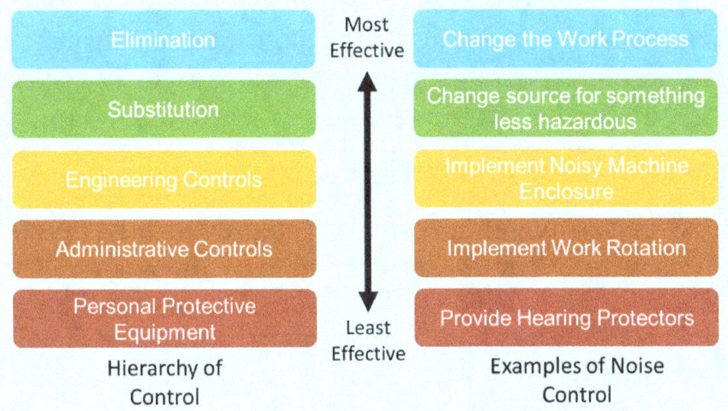

Fig. 4.4   Hierarchy of control.

Employers, self-employed persons, and principals are required to "eliminate foreseeable risk", which means the hazard is removed totally. If it is not "reasonably practicable" to eliminate the risk, employers, self-employed persons, or principals are to implement "reasonably practicable measures to minimise the risk" and safe work procedures to control the risk. The control measures identified in the RM Regulations can be found in Fig. 4.4, which is also known as the hierarchy of control. The control measures are defined as follows:

- "Substitution" means the replacement of any hazardous material, process, operation, equipment or device with less hazardous ones, and the risk of the hazard is reduced but not eliminated.
- "Engineering Controls" means the application of any scientific principle for the control of any workplace hazard; and includes the application of physical means or measures to any work process, equipment or work environment, such as the installation of any barrier, enclosure, guarding, inter-lock or ventilation system. Engineering control usually reduces the likelihood of the hazard causing injuries and ill health, but it can also reduce the potential severity.

- "Administrative Controls" means the implementation of any administrative requirement, which includes a permit-to-work (PTW) system. Administrative control is heavily dependent on the employee's compliance and careful implementation.
- "Personal Protective Equipment" (PPE) refers to equipment, such as helmets, safety boots, harnesses, and safety goggles, that a person wears to protect against hazards. PPE usually reduces severity but not the likelihood of an injury or ill health.

The hierarchy of control indicates that elimination is the most effective approach, while PPE is the least effective. This is because PPE only minimises the impact of hazards (i.e., severity) and is heavily reliant on the user always using the PPE correctly. In contrast, elimination removes the hazard and the need for any form of control. Substitution and engineering control are highly desirable, but organisations tend to over-rely on administrative controls and PPE to manage WSH risk because they are more easily implemented. It should be noted that multiple control measures should be implemented concurrently in case any of the control measures fail unexpectedly. For example, a worker working at height should be protected by barricades, but if the worker needs to reach over the barricade during their work, then a fall protection system (e.g., full body harness, lanyard with personal energy absorber, and anchorage), which is a form of PPE, should be used by the worker to reduce the risk of falling from height. Table 4.1 shows some examples for each type of control in the hierarchy of control.

ISO 31000:2018 uses the term risk treatment instead of risk control. Section 6.5.2 of ISO 31000:2018 highlights that risk treatment options are not mutually exclusive and can include the options stated in Table 4.2. As noted in Chap. 2, Haddon's ETM is also a useful guide for identifying possible risk controls.

Table 4.1 Examples of controls.

Type of control	Example
Elimination	• Using a long arm roller paint brush to eliminate the need to work on scaffolds • Designing windows that can be cleaned from indoors (e.g., can be turned inwards) to eliminate the need for workers to clean the windows from outside and at heights
Substitution	• Substituting flammable and potentially toxic solvent-based paints (i.e., may release flammable Volatile Organic Compounds) with water-based (also known as acrylic emulsions) paints • Substituting a noisy machine (>85 dBA) with a less noisy machine (~60 dBA)
Engineering control	• Barricades along open edges of a building • Ventilation to reduce the concentration of toxic chemicals in the atmosphere • Sensors to detect human motion in the vicinity of a machine and stop the machine when the sensors are triggered
Administrative control	• Work at heights training for workers working in mobile elevated work platforms • Permit to work for workers entering confined space • Warning signs at the bottom of a scaffold under erection to remind workers of the need to stay away
PPE	• Use of ear plugs by workers working in noisy area(s) • Use of eye protection for workers using cutting machines • Dust mask for workers cleaning dusty areas

Table 4.2 Risk treatment options described in ISO 31000:2018.

ISO 31000:2018 risk treatment options	Comments
Avoiding the risk by deciding not to start or continue with the activity that gives rise to the risk	Similar to elimination in the hierarchy of control, but instead of eliminating the hazard, the focus is on stopping work so as to prevent exposure to the hazard.
Taking or increasing the risk in order to pursue an opportunity	ISO 45001:2018 encourages organisations to proactively look for opportunities to improve WSH performance. However, due to the negative nature of accidents and ill health, most WSH risk assessments do not focus on opportunities.
Removing the risk source	This is the same as the elimination of the hazard.
Changing the likelihood	Most WSH risk controls change likelihood rather than consequences.
Changing the consequences	Reducing consequences of an incident can be achieved through reducing the amount of energy or substances stored and improving the resistance and resilience of individuals or organisations (see Chap. 2 on the Energy Transfer Model).
Sharing the risk (e.g., through contracts, buying insurance)	Also known as risk transfer, where specialist contractor or organisations and individuals with suitable competencies are selected to share the risk. When contractors are engaged, the principal who engaged the contractors are still expected to conduct RA.
Retaining the risk by an informed decision	This option is only viable if the risk is already ALARP. Organisations must be able to demonstrate ALARP and that the decision was an "informed decision".

A "safe work procedure" (SWP) is frequently seen as a type of administrative control measure, but the Technical Advisory for Working at Height (Workplace Safety and Health Council, 2008) describes SWP as "a set of systematic instructions on how work can be carried out safely. Arising from the risk assessment, a set of SWPs should be written for various jobs on site. The SWP provides (1) a step-by-step account of how jobs are to be executed, (2) who oversees these jobs, (3) what safety precautions must be taken (based on the risk assessment made earlier; *including emergency procedure*) and (4) what training is necessary for the workers doing these jobs. The (5) PTW system must be integrated with the SWP so that the supervisors are aware of the safety requirements and checks. The SWP must be (6) communicated to everyone involved in the job so that each is aware of the role they play in it".

Therefore, an SWP is not a control measure but a documentation of different control measures integrated with the work steps. Thus, for it to be effective, it must be detailed and must include the risk controls to be taken in the course of the work and during an emergency. In addition, SWPs must highlight suitable PPE that must be provided to the persons carrying out the work. Thus, the SWP is essentially a set of work instructions that includes risk controls identified in the RA.

There is no fixed format for SWPs, but they should be written such that they are easily comprehended by the people conducting the work. A useful reference is well-written assembly instructions for self-assembled furniture. These instructions include step-by-step guidance to buyers on how to put the furniture together as well as safety measures, such as putting a mat to protect the person assembling the furniture from injuries due to prolonged kneeling on a hard floor during assembly.

An RA is expected to be recorded, disseminated and reviewed. The documentation of the RA is necessary because it helps to ensure detailed analysis of the work and establishes accountability. RAs should be reviewed every three years, when there is an incident, a significant change to the work, or when relevant WSH information becomes available.

4.4 Overview of the risk management process

The Code of Practice on Workplace Safety and Health Risk Management (Workplace Safety and Health Council, 2022) describes RM processes that an organisation should implement (see Fig. 4.5). As can be seen, RA is a critical step of RM.

Fig. 4.5 Risk management process. (Adapted from (Workplace Safety and Health Council, 2022)).

4.4.1 Preparation

The RM process starts with the formation of an RM team, or RA team if the workplace is relatively small and has less variety of activities and hazards. An RM team consists of an RM leader (or champion) and multi-disciplinary team members who have knowledge of the different hazards and relevant controls.

The RM leader or champion must have attended suitable RM training and be able to formulate an RM plan customised to the needs of the organisation. An RM plan is essentially a plan to implement different controls so as to manage the WSH risk level of the workplace.

Another important responsibility of the RM leader is to select the relevant team members with the right RM training, relevant expertise, and suitable job responsibilities. The RM leader should have the ability to chair RA meetings effectively, i.e., they should have the ability to facilitate discussions, get members to contribute different opinions, and help the group arrive at agreed decisions efficiently. The RM leader also needs to ensure proper documentation and record retention. Some possible team members include management staff, site engineers, technicians, safety personnel, supervisors, operators, contractors, and suppliers. The roles and responsibilities of the team members must be clearly established, and only personnel trained in RA can be appointed as team leaders.

For large organisations and organisations with a wide range of activities, there will be a need for RA teams to be created. Each RA team will be focused on specific activities or areas in the workplace, and they will report to the RM team. The RA team leader will also be a member of the RM team. If the hazards and controls need specialist knowledge, or if the workplace does not have people versatile with RM or RA, it may be useful to engage consultants (e.g., engineers and technical specialists) to assist with the process. However, the consultants should be there to assist, and the RM or RA teams must still conduct the RM process and make the key decisions, such as the determination of risk levels and selection of controls.

The RM team will have to create an inventory of work activities and scope the RA. Scoping is essentially determining the boundaries for each RA. This can be based on the activities, location, type of personnel, material, etc. The scope of RAs will also be guided by the organisation's policy, objectives, targets, and programmes. For example, a developer may want to focus on hand and finger injuries arising from lifting work because, based on their experience, these are some of the most frequent types of injury. In this case, an RA team can be specially formed and scoped to investigate hand and finger injuries during crane lifting work. The RA will then form the basis for WSH management activities, such as a campaign to raise awareness on the prevention of hand and finger injuries during lifting.

During preparation, the RM or RA team will have to gather relevant information such as layout, workflow, manufacturer instructions, safety data sheets (for chemicals), etc. Different team members can assess different aspects of the collated information and then discuss them in meetings.

4.4.2 Risk assessment

After the preparation stage, an RA will have to be conducted. This stage includes hazard identification, risk evaluation, and risk control. During hazard identification, the RM or RA team will have to brainstorm for possible hazards. There are many hazard identification methods. One of the most basic is to review relevant hazards or safety issues documented in records such as WSH audit reports, inspection data, incident investigation reports, injury and illness records, occupational health surveillance reports, first aid logs, observation records, employee feedback or consultation records, and safety data sheets. These documents will supplement systematic hazard identification or RA techniques like Job Safety Analysis (JSA), Job Hazard Analysis (JHA), What-if Analysis, Failure Mode and Effect Analysis (FMEA), Hazard and Operability (HAZOP) Study, Fault Tree Analysis (FTA), and Event Tree Analysis (ETA) (see Table 4.3).

The wide range of hazard identification and RA techniques can be confusing, but in Singapore, most organisations use the

Table 4.3 Risk assessment techniques.

Technique	Description	Remarks
Job Safety Analysis (JSA) and Job Hazard Analysis (JHA)	JSA and JHA are very similar techniques, and both are conducted according to the following steps: • selecting the job to be analysed; • breaking the job down into a sequence of steps; • identifying potential hazards; • determining preventive measures to overcome these hazards. Some JSAs require the risk level to be determined, while JHAs usually do not require risk levels to be determined. This is the key difference between the two techniques, but many variants exist.	JSA and JHA are relatively simple to use. They are usually structured based on a table similar to the table recommended in the RMCP. However, JHA does not require risk levels to be identified. They are suitable if the focus is on an activity, process or task that has specific steps. They are the most commonly used techniques in the construction industry.
What-if Analysis	A What-if Analysis consists of structured brainstorming to determine what can go wrong in a given scenario. The person performing the analysis then judges the likelihood of things going wrong and considers the consequences. What-if Analysis can be applied at virtually any point in the evaluation process and is essentially	The what-if questions should be generated before the brainstorming session, and a group of experts must be involved in the analysis. Guided by a detailed template, the What-if Analysis is comparatively unstructured.

(Continued)

Table 4.3 (*Continued*)

Technique	Description	Remarks
	a brainstorming session to identify hazards and assess their risk levels. People involved should be experienced, and the facilitator must be able to guide the group. The discussion will provide informed judgements about the acceptability of risks. A course of action can be outlined for unacceptable risks (adapted from https://www.acs.org/).	
Failure Mode and Effect Analysis (FMEA)	The FMEA is also a template-based approach, and it has a long history in reliability engineering. FMEA typically divides a piece of machinery or equipment into subsystems that can be assessed effectively. The process involves (1) Identification of the component and parent system, (2) Failure mode and cause of failure, (3) Effect of the failure on the sub-system or system, and (4) Method of detection and diagnostic aids available. FMEA is quantitative in nature, and it estimates the risk levels based on failure and error data collected by plants and manufacturers.	FMEA is commonly used in the manufacturing industry. Its quantitative nature makes it feasible only when the data is readily available. Nevertheless, it is still possible to use it based on qualitative assessments.

Hazard and Operability (HAZOP)	HAZOP is an established approach for reviewing chemical process designs. It is a systematic search for hazards defined as deviations within parameters that may have dangerous consequences. In the process industry, these deviations concern process parameters such as flow, temperature, pressure, etc. (adapted from www.hse.gov.uk).	During a HAZOP study, the team will typically spend a few days to a few weeks evaluating a plant's piping and instrumentation diagrams to be constructed. The facilitator will use a set of guidewords (e.g., more, less, high, low, fast, slow) to match the process parameters to brainstorm what are the possible effects of these deviations. Commonly used in industries like oil and gas, chemical, and energy-related.
Fault Tree Analysis (FTA) and Event Tree Analysis (ETA)	A fault tree is a diagram that displays the logical interrelationship between the basic causes of the hazard. The complexity of an FTA depends on the system being assessed. Complex FTA involves the use of Boolean algebra to represent various failure states (adapted from www.hse.gov.uk). An ETA is a forward, bottom-up, logical modelling technique for both success and failure that explores responses through a single	FTA and ETA are commonly used in the nuclear and process industry to analyse complex engineering systems. Human errors can also be included to determine the reliability of these complex systems. FTA and ETA are quantitative in nature but can be adapted to be used qualitatively.

(Continued)

Table 4.3   (*Continued*)

Technique	Description	Remarks
	initiating event and lays a path for assessing probabilities of the outcomes and overall system analysis. This analysis technique is used to analyse the effects of functioning or failed systems, given that an event has occurred (adapted from https://en.wikipedia.org/wiki/Event_ tree_analysis).	
Bowtie Analysis	The Bowtie Analysis is an RA method that can be used to analyse and communicate how high-risk scenarios develop. A bowtie gives a visual summary of all plausible risk scenarios concerning a certain hazard that could occur. By identifying control measures, the bowtie displays what a company does to control those scenarios. Control measures are also known as barriers. They are elements we have in place to ensure the hazards we deal with are kept in a wanted state. Barriers can be systems, regulations, design aspects, and so on (adapted from Bowtie Methodology Manual by CGE).	The bowtie is a qualitative method that relies on a set of symbols to summarise a series of hazards and controls in diagrammatic form. The bowtie requires specialised software to be implemented effectively, as the number of symbols and diagrams will grow quickly. Each bowtie diagram is a good summary of the hazards and controls a particular activity has.

approach recommended in the Risk Management Code of Practice (RMCP), which is essentially an activity-based RA (similar to JSA and JHA) or a trade-based RA. The activity-based RA and trade-based RA will be elaborated on later in this chapter. In addition, if the workplace already exists, an inspection or walkabout can be conducted as part of the RA. Furthermore, during these walkabouts, checklists can be used to help RA teams identify hazards and observations, and interviews can also be conducted to obtain valuable information about how existing controls are performing. In general, the following categories of hazards should be considered:

- physical (e.g., fire, noise, ergonomics, heat, radiation);
- mechanical (e.g., moving parts, rotating parts);
- electrical (e.g., voltage, current, static charge, magnetic fields);
- chemical (e.g., flammables, toxics, corrosives, reactive materials);
- biological (e.g., blood-borne pathogens, virus);
- psychosocial (e.g., stress, fatigue).

Human and behavioural factors are important aspects that must be considered during RA. Human factors literature (e.g., (Health and Safety Executive, 1999)) identified two broad types of human failures: errors and violations (see Fig. 4.6). The key difference between errors and violations is that violations are deliberate, and errors are unintentional. However, both involve an individual not acting in accordance with accepted standards, rules or procedures.

Skill-based errors are further classified into slips of action and lapses of memory. Slips of action, or slips, happen when an individual performing a very familiar task performs the wrong action. Using driving as an example, slips can involve stepping on the brake too late, not checking the blind spot, stepping onto the accelerator too hard, making a wrong turn, and stepping

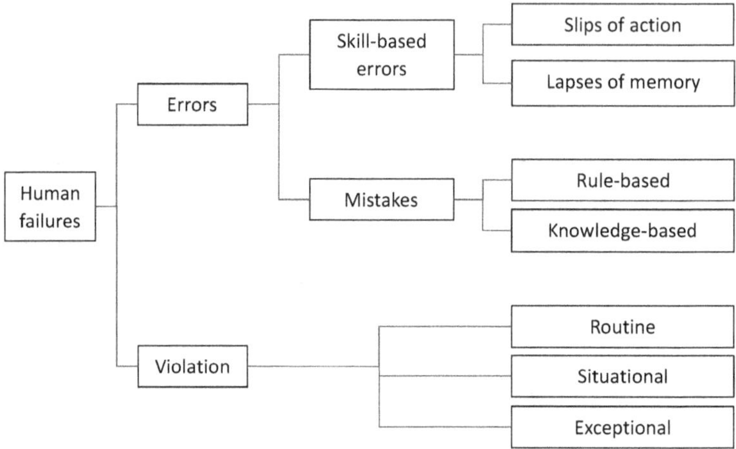

Fig. 4.6 Types of human failures.

onto the accelerator when the intention is to brake. Lapses of memory, or lapses, happen when an individual performing a very familiar task forgets to do an action, gets lost during a series of actions, or forgets the purpose of the actions.

Mistakes involve the failure of mental processes during planning, assessing, forming intentions, and judging conse-quences. Rule-based mistakes occur when we use familiar rules that are inappropriate for the situation. For example, an engineer inspecting a faulty crane reads the manufacturer's instructions for one of the components and assumes that the instructions apply to the other components. Hence, he fails to replace a safety sensor, which results in a crane accident later. Knowledge-based mistakes occur in unfamiliar situations when an individual has to diagnose the situation to select the right action. An example of a knowledge-based mistake is when a construction work crew detects cracks on the supporting struc-tures during pouring of concrete — and as the company does not have a pre-determined monitoring and evacuation procedure, the site personnel discuss the appropriate actions and misdiagnose the cracks, and while doing so, fail to evacuate the workers, resulting in a major accident.

Some of the key factors influencing the occurrence of human errors include presence of work environment stressors (e.g., heat, poor lighting, noise, and confined space), excessive task demands (e.g., high workload, unrealistic expectations on alertness, repetitive and monotonous tasks, excessive distractions), social and organisational stressors (e.g., insufficient human resource, demanding shift arrangement, conflicts between colleagues, peer-pressure, and work bullies), individual stressors (e.g., lack of knowledge, fatigue, non-work problems, and mis-use of alcohol or drugs), and equipment stressors (e.g., poorly designed user interface and unclear instructions). These factors must be considered during the RA.

Violations are another aspect of human failures that must be considered during RA. Many accidents occur due to violations, and it is necessary for every RA to consider the tendency for workers to take shortcuts or violate SWPs, safety rules, and risk controls. Some possible violations include bypassing safety checks, disabling machine safety features, not using or misusing PPE, and unauthorised use or misuse of equipment. Violations can be routine, situational or exceptional. Routine violations are an indication that breaking safety rules is the norm in an organisation and reflects the detrimental state of safety culture in that organisation. Situational violations happen when there are certain stressors arising from the individual's work environment, equipment, job, social context, or organisation that appear to justify the violations. Exceptional violations happen only during emergencies or rare events, where emergency procedures or SWPs are violated based on the wrong belief that the benefits of the violation outweigh the cost of following them.

When considering human factors during RA, the RA team must also consider individual health risk factors, including medical health issues, smoking, and alcohol misuse.

The risk matrix in Fig. 4.7, together with the severity level and likelihood levels in Tables 4.4 and 4.5, respectively, are described in the RMCP. They are used to evaluate hazards and assign a risk priority number (RPN). Once the RPN is determined, the risk level can then be determined, and the RM team can then refer to Table 4.6 for the recommended actions.

Figure 4.7 and Tables 4.4, 4.5 and 4.6, taken from the RMCP, are guidelines, but they represent industry norms and generally accepted risk tolerance levels. Thus, the recommended risk matrix in the RMCP and other national codes of practice should be considered the default risk matrix for the relevant countries unless the organisation has a better alternative. Nevertheless, organisations can create their own risk matrix, severity, and likelihood levels, and recommended actions and calibrate them to reflect their risk appetite and tolerability. However, any deviation from the guidelines provided in the RMCP should be conservative, i.e., the organisation's risk appetite should be less than or equal to the industry norm. One way to calibrate risk matrices and levels is to identify common hazards that most employees of the organisation are familiar with. Next, sample

SEVERITY	LIKELIHOOD				
	1 Rare	2 Remote	3 Occasional	4 Frequent	5 Almost Certain
5 CATASTROPHIC	5 MEDIUM	10 MEDIUM	15 HIGH	20 HIGH	25 HIGH
4 MAJOR	4 MEDIUM	8 MEDIUM	12 MEDIUM	16 HIGH	20 HIGH
3 MODERATE	3 LOW	6 MEDIUM	9 MEDIUM	12 MEDIUM	15 HIGH
2 MINOR	2 LOW	4 MEDIUM	6 MEDIUM	8 MEDIUM	10 MEDIUM
1 NEGLIGIBLE	1 LOW	2 LOW	3 LOW	4 MEDIUM	5 MEDIUM

Fig. 4.7 Risk matrix. (Adapted from (Workplace Safety and Health Council, 2022)).

Table 4.4   Severity level (Workplace Safety and Health Council, 2022).

Level	Severity	Description
5	Catastrophic	Death, fatal occupational disease or exposure, or multiple major injuries
4	Major	Serious injuries, serious occupational diseases, or exposure (includes amputations, major fractures, multiple injuries, occupational cancers, diagnosed mental illnesses, acute poisoning, disabilities, and noise-induced hearing loss)
3	Moderate	Injury or ill-health (including mental well-being) requiring medical treatment (includes lacerations, burns, sprains, minor fractures, psychosocial stress, dermatitis, and work-related musculoskeletal disorders)
2	Minor	Injury or ill-health (including mental well-being) requiring first-aid only (includes minor cuts and bruises, irritation, ill-health with temporary discomfort, and fatigue)
1	Negligible	Negligible injury

Table 4.5   Likelihood level (Workplace Safety and Health Council, 2022).

Level	Likelihood	Description
1	Rare	Not expected to occur but still possible
2	Remote	Not likely to occur under normal circumstances
3	Occasional	Possible or known to occur
4	Frequent	Common occurrence
5	Almost Certain	Continual or repeating experience

employees who are reflective of the population to assess those hazards individually. Subsequently, the RPNs generated by the sampled employees can be used to assess if the definitions in Tables 4.4 and 4.5 are suitable. The higher the RPN, the more

Table 4.6   Recommended actions for risk levels (Workplace Safety and Health Council, 2022).

Risk level	Risk acceptability	Recommended actions
Low	Acceptable	• No additional risk control measures may be needed. • Frequent review and monitoring of hazards are required to ensure that the risk level assigned is accurate and does not increase over time.
Medium	Tolerable	• A careful evaluation of the hazards should be carried out to ensure that the risk level is reduced to ALARP within a defined period. Interim risk control measures, such as administrative controls or PPE, may be implemented while longer-term measures are being established. • Management attention is required.
High	Not acceptable	• High-Risk level must be reduced to at least Medium Risk before work starts. • There should not be any interim risk control measures. Risk control measures should not be overly dependent on PPE. • If practicable, the hazard should be eliminated before work starts. • Management review is required before work starts.

attention and control are required. The hazards assessed to be in the top right corner (where the RPN = 15 to 25) need additional risk controls before work can be conducted. Controls can be identified based on the hierarchy of control and Haddon's countermeasures. The diagonal area of the matrix (where RPN = 4 to 12) indicates that the risk level is tolerable (medium), but additional control measures should be implemented whenever reasonable and practicable. The lower left zone (where RPN = 1 to 3) indicates that the hazard is low-risk and generally acceptable, but practicable controls that can push the risk lower should still be implemented. The different zones are aligned with the HSE Framework for Tolerability of Risk (Fig. 4.4).

An RA is meant to be a live document. As highlighted in the WSH (RM) Regulations, the employer, self-employed and principal are required to review the RA where there is a "significant change in work practices or procedures" (see Regulation 7(2)(b)). The RMCP, Clause 4.2.1.12, also indicates that "upon any accident, incident, near miss or dangerous occurrence" and "when new information on WSH risks is made known", the RA must be reviewed and, if necessary, revised. An RA review is similar to an RA, but instead of starting from scratch, the RA Team will review existing RAs based on recent incidents, changes, and/or additional information.

Even though RA is a systematic process that should lead to "consistent and reliable results" (RMCP Clause 6.2.5), it is only useful if the employer, self-employed or principal and its RA Team are committed to safety and conduct the RA carefully, meticulously and seriously. If the employer, self-employed or principal and its RA Team treat RA as a "paper exercise" only to satisfy requirements, the RA will not be effective and will not have significant benefits on WSH. Furthermore, the RA Team is not meant to work independently. They should keep the RM Team and management updated and seek their approval and support whenever necessary.

Another important consideration during an RA is highlighted in Clause 4.3.1.4.4 of OHSAS18002:2008 Occupational Health and Safety Management Systems — Guidelines for the Implementation of OHSAS 18001:2007 (British Standards Institution, 2008), which states that, "Hazards that could cause harm to large numbers of persons should be given careful consideration even when it is less likely for such severe consequences to occur". This indicates that the RA team should always be cautious and give more weight to consequences so as to effectively prevent rare major accidents. Major accidents, a "black swan" event, are unlikely events in terms of occurrence frequency, but they have very severe consequences.

The RA team must be mindful that they do not have full knowledge of hazards, risks (especially likelihood), and effectiveness of their controls. Therefore, it is critical for RA teams to develop monitoring procedures and continuously update the RA whenever there are changes to work practices, incidents, near hits, and when new WSH information is available. An example is the Nicoll Highway collapse, where information about waler beam buckling and abnormal instrumentation readings were not carefully evaluated as part of an RA review. The project team continued to assume that a tunnel collapse was not possible, possibly because major tunnel collapses had never occurred in Singapore prior to the accident.

It must be noted that black swan events typically appear to be rare or extremely unlikely to organisations because of lack of information or knowledge or failure to systematically consider available information. Thus, by requiring a structured and continuous RA process, organisations will be less likely to miss critical WSH information and can guard against major accidents.

4.4.3 Implementation, record-keeping and review

To facilitate implementation, there should be an RM plan at the end of the RA to ensure that all stakeholders are aware of the key hazards, their risk levels, and the corresponding control. The RM plan should define the scope of the RM, the company job functions, the RA methodology, the risk control measures, and the schedule or programme for implementing the RM plan. Most importantly, the RM plan should contain the actions that are required to implement the different controls identified during the RA. Frequently, stopgap measures (usually lower on the hierarchy of control, i.e., administrative controls and PPE) need to be implemented first because more effective measures, e.g., installing an engineering control, need time for implementation. These details need to be spelt out in the RM plan, such that the risk level during operations is at most tolerable or medium. The plan should clearly stipulate:

- what controls have to be implemented;
- how to implement the controls;
- who are to implement the controls;
- where to implement the controls;
- when to implement the controls.

During implementation, it is important for organisations to ensure buy-in from supervisors and workers through communication and consultation. In terms of communication, it is critical to ensure that the nature of the hazards, risk levels and controls are explained. The relevant safe work procedures should be communicated in a suitable format to the relevant stakeholders, e.g., in pictures or comics to overcome possible communication barriers. The controls will impact the work environment, equipment, work procedures, and personnel. Thus, the impact of the changes should be clearly communicated before implementation.

For example, the RA team for the installation of a drywall along the edge of a mezzanine floor in a building was in a rush to finish the RA. They identified the hazard of falling over the edge of the mezzanine floor (3 m above the ground floor) in a new workshop and noted that due to the lack of height clearance, lanyards and personal energy absorbers will not be suitable to protect the workers. Thus, they recommended the use of safety or anti-fall nets to be installed along the perimeter of the mezzanine floor. However, the nets made it infeasible for the scissors lifts to be used to assist in the installation of the drywall. Workers familiar with the work should have been able to identify the problem easily but were not consulted because the RA team conducted the RA in a rush. This resulted in delays in the project as a new RA had to be conducted and the installed safety net had to be removed.

To facilitate communications, the organisation should clearly establish the different channels of communication, including WSH committee meetings, feedback sessions, small group meetings (e.g., tool-box meetings and shift handover meetings), one-on-one discussions, email or online messaging, telephone calls, notice boards, and bulletins. The stakeholders that must be involved in the development and communication of the RM plan can include senior management, supervisors, subject matter experts, workers, consultants, contractors, and suppliers. The effectiveness of the controls and risk of hazards must be monitored during implementation. The monitoring process must be carefully considered during the RA, where the RA team, with inputs from relevant experts, must consider possible sources of uncertainty or inaccuracy that could render the RA unreliable. These uncertainties can be due to a lack of information (during the RA) about the work, workers, environment, equipment, stakeholders, and risk controls. The monitoring system can be based on inspections, sensors or instruments, such as CCTV or other technology, but it should be suitable and proportionate to the work situation.

As highlighted earlier, RA reviews should be conducted at least once every three years, whenever there are significant changes to the work, when there are incidents, near-hits or dangerous occurrences, or when new information on WSH risks is made known. The RM plan should also clearly define the communication, consultation, and review activities.

4.5 Common errors in risk management

Despite its importance, many RM and RA processes are saddled with errors during actual implementation. Table 4.7 summarises the common errors identified in the document "E-Fact 32 — Common errors in the risk assessment process" (European Agency for Safety and Health at Work, 2008), adapted based on the author's experience.

Besides the common errors described in Table 4.7, it must be noted that risk perception is affected by many cognitive biases (based on (Dobelli, 2013)) such as:

- Base rate neglect: A disregard of fundamental frequencies of different types of accidents and illnesses. During RA, organisations fail to consider the overall accident frequencies in the industry and assign unrealistic likelihoods to their own workplaces.
- Over-confidence: When individuals are asked to rate their competency or likelihood of getting into an accident, most will over-rate their competency in relation to others and assume that they will not get into an accident.
- Availability bias: Most people do not get into an accident or illness daily. Thus, based on their experience, the most available memory is those of non-accidents or days with no illness. This can lead to developing a mindset biased towards the assumption that accidents and illness will not happen in their workplace.
- Survivorship bias: Similar to the above biases, we tend to focus on success or "non-failure". As Dobelli (2013) puts it, "People systematically overestimate their chances of success. Guard against it by frequently visiting the graves of once-promising projects, investments and careers. It is a sad walk, but one that should clear your mind."

Table 4.7 Common errors in risk management.

RM stage	Common error	Remarks
Preparation	• Not involving a team of people in the assessment	• RM and RA should be a team effort, where different opinions and ideas are collected to ensure effectiveness (e.g., comprehensiveness of hazards identified, representativeness of risk levels assigned, and proportionality and suitability of risk controls selected).
	• Not including people with the right expertise and experience	• Expertise can include technical or specialist knowledge of the work process, environment, material, and equipment used. • Experience refers to the practical experience of carrying out the work; experienced frontline workers and/or supervisors should be involved.
	• Not designating the RA to a person who is competent	• RA process requires competencies such as facilitation skills, knowledge of risk assessment techniques, and the ability to communicate with different stakeholders.
	• Involving experts (including consultants) in the RA process who are not familiar with the organisation	• It is useful to involve experts with technical expertise, but the RM/RA team needs to orientate them to the RM/RA process so that the experts understand the relevant WSH policies and procedures, the RM/RA method used, the scope of their engagement, the role of the RM/RA team, and the resources available to the expert.

		• It is important that the experts work with the RM/RA team to conduct the RA; if the experts conduct the RA without any involvement from the RM/RA team, the RA process may become impractical, and there may be limited buy-ins, therefore causing problems during implementation.
	• Not allocating enough time and resources to the RM/RA process	• RM/RA process is a planning process that should be given enough emphasis, resources and support from management; management frequently underestimates the time and resources needed for a thorough RM/RA process.
	• Not collecting enough background or site-specific information	• RM/RA team may miss required information during preparation, e.g., site-specific characteristics (e.g., traffic condition and community profile), information on specific types of equipment to be used (including the age of equipment), and documentation on past incidents and ill-health.
Risk Assessment: Hazard Identification	• Overlooking possible hazards and incidents	• RM/RA team may miss hazards that they are not familiar with, such as health risks (e.g., psychosocial issues and chronic diseases). • RM/RA team may assume that hazards and incidents that had never happened before will never happen in the future, therefore causing hazards that are of low likelihood but high severity to be overlooked; it is important for hazard identification to be comprehensive and not screen out possible hazards too early; the filtering and prioritisation should only be done during risk evaluation.

(Continued)

Table 4.7 (*Continued*)

RM stage	Common error	Remarks
		• Inappropriate use of checklists can cause hazards to be overlooked; checklists are usually the first steps to hazard identification, and RM/RA teams will need to go beyond the checklists to ensure that the hazards identified are comprehensive. • Hazards can be overlooked if the work process is not carefully documented; non-routine tasks and activities like maintenance, repair and remedial or corrective actions when foreseeable errors were made need to be included as part of the work and task inventory. • Hazards arising due to interactions between different work activities due to their proximity, sequence or dependencies can be overlooked because the RM/RA team was focusing on each of the activities separately.
	• Overlooking other people exposed to the hazards	• Hazards may be identified, but other people exposed to the hazards, such as other employees not directly involved in the work, subcontractors, visitors, and members of the public, may not be adequately identified.
	• Overlooking the vulnerability of specific groups of people exposed to the hazards	• The RM/RA team typically focuses on the "average worker" exposed to a hazard, but specific groups of workers, e.g., older workers, pregnant or workers with disability, may be more susceptible to certain hazards and may require additional control measures.

	• Overlooking specific characteristics of the work being assessed that differ from the past	• The RM/RA team may be over-relying on past RAs or past experience; it is important to customise the RA by considering the specific characteristics of the current work, e.g., location, work methods, workers profile, equipment, material and structure, and to consider the changes made since the last RA and not assume that everything is business as usual.
	• Assuming that a hazard will be looked into by others	• RM/RA team assumes that another RM/RA team, engineering team, or other relevant teams is looking into a hazard and removes the hazard from the team's scope of work.
Risk Assessment: Risk Evaluation	• Not considering long-term consequences	• Long-term consequences, especially health-related consequences, may not be assessed.
	• Lack of consistent description of the hazard and possible incident	• The RM/RA team may not develop a coherent description of the hazard and possible incident so that the risk levels considered by different RM/RA team members are based on the same scenario. • A useful approach is to assess all hazards based on the concept of "worst credible scenario", where a reasonable estimation of the worst-case scenario is used to ensure consistency across the risk evaluation.

(*Continued*)

Table 4.7   (*Continued*)

RM stage	Common error	Remarks
	• Not considering the reliability and effectiveness of risk control measures when assigning a risk level	• When considering the severity and likelihood of a hazard and its potential consequences, the RM/RA team might neglect to consider the likelihood of the risk control not performing as intended due to issues such as equipment failure, human error, and violations.
	• Reducing severity, when likelihood should be reduced	• When additional controls are implemented, the severity should be reduced only if the controls are reliable in reducing the energy or amount of hazardous substances that a worker will contact during an incident. • Using the "worst credible scenario" rule, severity is less likely to be reduced when compared with likelihood.
	• Not doing validity checks on the risk levels assigned	• RM/RA team should sort all hazards based on the risk level and check if the risk levels assigned make sense when compared across the hazards. • Another approach is to compare with benchmark risk levels that most people can relate to, e.g., risk level of crossing the road, driving, and taking public transport.

Risk Assessment: Risk Control	• Not taking the hierarchy of control into consideration	• RM/RA team might think that elimination is not possible and did not even brainstorm on possible opportunities to eliminate hazards. • RM/RA team might over-rely on administrative controls (e.g., training, warning signs, and briefings) and PPE and not strive to identify substitution and engineering controls.
	• Assume that transferring the risk means no risk	• Risk transfer is better known as risk sharing; when contractors or consultants are brought in for a specific scope of work, the principal engaging the contractor or consultant will have to work with them to manage WSH risks and not assume that they have transferred all their risk and no longer have any WSH duties.
	• Not prioritising the controls, clearly assigning employees overseeing the implementation of the controls, and stipulating due dates	• Due to the wide range of hazards identified, it is not possible to implement all the additional controls immediately; thus, it is important to prioritise the controls based on the inherent risk levels (risk level before additional controls) and clearly identify the employees to ensure implementation of the risk controls with strict due dates. • Special attention should be given to controls for hazards with high severity but low likelihood; these hazards can cause catastrophic consequences, and their prevention is dependent on reliable controls that are closely monitored. • The due dates for implementation should be tied to the start date of the relevant task or activity.

(Continued)

Table 4.7   (*Continued*)

RM stage	Common error	Remarks
Implementation and Review, and Communication	• Not involving or communicating with workers	• Implementation of risk controls is heavily dependent on buy-in and cooperation from workers; it is important that there is communication and consultation with workers on the risk controls and give them opportunities to feedback on the controls. • Workers must feel safe to work, and communication affects their risk perception; therefore, management must continue to assure workers of the effectiveness of the risk controls and encourage hazard reporting if any of the controls is in doubt.
	• Not reviewing the RA when necessary	• A RA review must be conducted at least once every three years, but when there are significant changes to the work, when incidents arise, and when new WSH information becomes available, the RA should be reviewed. • Organisations and individuals may assume that certain work changes, incidents and WSH information are not significant enough to trigger an RA review; on the contrary, RA reviews should be expected because RAs are frequently conducted prior to the beginning of work, so it is frequently inaccurate; thus as more information becomes available, RA reviews must be conducted.

		• Management should make RA reviews a norm and a natural part of work, especially highly dynamic and project-based work, e.g., construction, and shipbuilding and ship repair work.
	• Not monitoring hazards, risk level, and risk controls	• During implementation, the organisation must continue to monitor the hazards, risk level, and risk controls to ensure that the risk level does not become unacceptable unknowingly.
	• Not implementing the risk controls in the RA	• Organisations may treat the RA as a "mere paper exercise" for the sake of satisfying regulatory requirements (i.e., surface compliance), and the risk controls are not truly implemented in the workplace. • Organisations with poor safety culture would typically have such behaviours; it is up to management to proactively check that RA is given emphasis and attention so that frontline employees are motivated to implement the risk controls.
Keeping Record	• Not recording RA	• RA is meant to be a deliberate and structured approach, which prevents organisations and individuals from making intuitive judgements that are error prone and risky; the process of having the RA recorded is meant to facilitate the deliberate and structured approach. • Information regarding hazards, risk levels, and risk controls should also be carefully recorded to ensure that critical WSH-related information is considered.

(Continued)

Table 4.7   (*Continued*)

RM stage	Common error	Remarks
		• Recording also helps to ensure that information is clearly communicated to relevant personnel, provides the basis for RA reviews, facilitates tracking of the status of implementation, and facilitates demonstration of due diligence. • Larger organisations with multiple RAs should consider using RM software to help them with RM and RA records; the RM software should help to structure the different RAs to minimise repetitive information from being documented in different records; and the software should facilitate cross-referencing and updating of documents systematically.

To deal with these biases, it is important to communicate WSH information, statistics, and case studies to the managers, supervisors and workers. Awareness of these cognitive biases will also help to reduce their potential impact.

Another fundamental problem common in organisations is the failure to implement the RA due to a disproportionate focus on productivity and poor communication. These issues are related to the safety culture of the organisation. Some organisations are over-focused on the documentation needed in RM. They think that having the set of documents means that they are protected from possible prosecution by the authorities, rendering RM or RA a mere paper exercise. However, paperwork without implementation is still poor management — the paperwork takes up time without adding value. If accidents happen, the organisation will still be heavily penalised.

Communication should always be focused on the receiver's preferences and suitability for the mode of communication. For example, the use of pictures and videos is much more effective for workers and supervisors than lengthy RA documents. In industries where workers of different nationalities are present, the clarity of communication needs to be investigated carefully. The person conducting the briefing must prepare beforehand, conduct the briefing clearly, and check if the receivers of the information have understood the information.

4.6 Systems thinking and risk management

Lee and Green (2015) highlighted how systems thinking concepts can be used to improve enterprise RM. Since the processes and goals of RM are similar across different fields, the insights from Lee and Green (2015) are applicable to WSH RM. The following highlights some useful insights adapted from Lee and Green (2015).

First, systems thinking emphasises the whole and not single components. During RM processes, a divide-and-conquer approach is taken to structure and scope different RAs. This may inadvertently cause hazards that arise from the interactions between different system components and processes to be missed. Therefore, management must ensure that WSH managers or risk managers are assigned to maintain a holistic view of the RM processes to check that interfaces, interdependencies, and "grey areas" are being studied by specific RM/RA teams.

Second, systems thinking recognises that mental models and organisational culture are critical components of any system. To ensure effective RM, the management must recognise that risk is a human construct, and it is important to understand the mental models and organisational culture that influence risk perceptions. This is where risk communication must consider the mental models of stakeholders and organisational culture and plan and target the risk communication activities accordingly.

Third, it must be acknowledged that the overall system objectives (e.g., zero accident) may conflict with individuals' goals or other system objectives. This is where management must consciously identify such conflicts or the resulting resistance to assure the effectiveness of RM processes.

Lastly, double-loop learning (similar to the concept of underlying factors in Chap. 3) is more effective in the continuous improvement of WSH RM than single-loop learning. In the context of WSH RM, single-loop learning is about adjusting WSH objectives and approaches by comparing the current performance with established objectives. This can include the objective of implementing a series of risk controls and comparing that with the actual implementation on the ground. If the implementation objective is not achieved, management can continue to increase enforcement or communication. Such single-loop learning is the typical control loop used in most organisations. Double-loop learning is focused on evaluating the underlying assumptions related to WSH RM when comparing WSH performance against objectives. Using the same example, when risk controls are not implemented as planned, management should seek to understand the underlying mindset, cultural, systemic, or leadership issues that could be hindering the implementation. These systems thinking management principles are useful guidelines for managers overseeing RM.

4.7 Summary

RM is not unique to WSH management, and it can be found in different management areas. RM is the cornerstone of WSH management, and any workplace seeking to prevent accidents and ill-health must understand RM and implement it effectively. The RM process is a structured process of preparation, RA, implementation, review recording, and communication. It helps organisations to proactively reduce risk. However, there are several common errors in implementing RM, and the management must be aware of them. The chapter also described how systems thinking's focus on the holistic and a fundamental understanding of system behaviour provide useful management guidelines for all WSH RM efforts.

Review questions

1. Why is RA the cornerstone of the WSH Act?
2. What are the factors that affect risk perception?
3. Describe the roles of the following people in RM: Employer, Manager, Human Resource Manager, RM and RA Leaders, and Employees.
4. When can an organisation stop monitoring a specific hazard?

5. Describe six categories of hazards.
6. Describe three examples of human and cultural factors that should be considered in an RA.
7. Using the recommended 5 × 5 matrix, what are some of the high-risk, medium-risk and low-risk hazards in your industry? How do you justify your answer?
8. What do you think are some of the possible problems in implementing the RMCP in a small workplace?
9. What are some of the common errors in WSH RM?
10. What are some of the systems thinking concepts that can help to improve RM?

References

British Standards Institution. (2008). OHSAS18002:2008 Occupational health and safety management systems — Guidelines for the implementation of OHSAS 18001:2007. London: British Standards Institution.

British Standards Institution. (2009). PD ISO Guide 73:2009 Risk management — Vocabulary. London: British Standards Institution.

British Standards Institution. (2018). BS ISO 31000:2018 Risk management — Guidelines. London: British Standards Institution.

Committee on Safety and Health at Work. (1972). Safety and health at work: Report of the committee, 1970–72 (Roben's Report). London: H.M. Stationery Office. http://www.mineaccidents.com.au/uploads/robens-report-original.pdf (accessed: 24 June 2020).

Dobelli, R. (2013). *The art of thinking clearly: Better thinking, better decisions*. London: Hodder and Stoughton.

European Agency for Safety and Health at Work. (2008). E-Fact 32 — Common errors in the risk assessment process. https://osha.europa.eu/en/publications/e-facts/e-fact32/view (accessed: 24 February 2020).

Health and Safety Executive. (1999). *HSG 48 Reducing error and influencing behaviour*. 2nd edition. London: The Stationery Office.

Health and Safety Executive. (2001). *Reducing risks, protecting people: HSE decision-making process.* London: The Stationery Office.

Lee, L. S. and Green, E. (2015). Systems thinking and its implications in enterprise risk management. *Journal of Information Systems*, **29**(2), 195–210.

Workplace Safety and Health Council. (2008). Technical advisory for working at height. Singapore: Workplace Safety and Health Council.

Workplace Safety and Health Council. (2022). Code of practice on workplace safety and health (WSH) risk management. Singapore: Workplace Safety and Health Council.

Chapter 5

Design for Safety

The Event Causation Technique

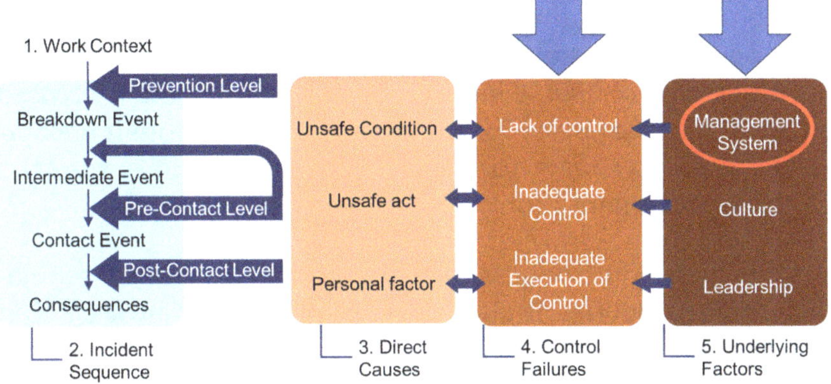

Design for Safety (DfS) is essentially a design risk assessment. It is an important aspect of the management system (see the Event Causation Technique (ECT) diagram above). The key outcomes of the DfS process are the design-related controls to prevent incidents downstream.

5.1 Introduction

Design for Safety (DfS) promotes early consideration of safety and health hazards during the design phase of a project, workplace or product. With early intervention, hazards can be more effectively eliminated or controlled, leading to safer products, workplaces, and production and construction processes. This chapter focuses on how DfS is practised in the built environment, but the processes and concepts also apply to other industries.

DfS is practised in the built environment of many countries, including Australia, the United Kingdom (UK), and Singapore. In Singapore, the Ministry of Manpower (MOM) enacted the Workplace Safety and Health (Design for Safety) Regulations 2015 (DfS Regulations) in July 2015, which were implemented from August 2016 onwards. Before the DfS Regulations, contractors were the main party involved in workplace safety and health (WSH). DfS imposes safety and health duties on the developers and designers, requiring them to conduct design risk assessments (RAs). Even though the DfS process is focused on WSH, it is also useful for the project team because it encourages the project team to assess risks upstream to ensure the effective implementation of the project downstream.

This chapter introduces the concept of DfS, relevant regulations, and guidelines. It also provides examples of how DfS can improve WSH throughout a structure's lifecycle and discusses its challenges and success factors for DfS.

5.2 Defining Design for Safety

According to numerous studies, many fatalities in the construction industry can be attributed to design decisions or lack of planning (Behm, 2005; Workplace Safety and Health Council, 2015). Thus, DfS was introduced to minimise the risk of accidents and ill health by considering hazards during the upstream design phases of a construction project (Gambatese *et al.*, 2008). DfS is also known as prevention through design (López-Arquillos *et al.*, 2015), safe design (Safe Work Australia, 2018), and Construction (Design and Management) (CDM) (Health and Safety Executive, 2015).

DfS is defined as: "The practice of anticipating and 'designing out' potential occupational safety and health hazards and risks associated with new processes, structures, equipment, or tools, and organising work, such that it takes into consideration the construction, maintenance, decommissioning, and disposal or recycling of waste material, and recognising the business and social benefits of doing so." (Schulte *et al.*, 2008, p. 115). Essentially, DfS is the RA of a design to identify the hazards that can pose safety and health risks at different development stages of a workplace, building or facility so that the hazard can be eliminated or mitigated through options such as design changes, risk control measures, and the sharing of information.

The DfS process requires the involvement of all stakeholders. The construction industry's stakeholders include the client or developer, the designers, and the contractors. However, one important stakeholder not mentioned in the DfS Regulations is facilities management, which holds important WSH knowledge of maintenance and operation. Similarly, in other industries, the owner or client usually engages designers and contractors to design, manufacture, or construct an asset, equipment, machinery or product. This asset, equipment, machinery or product can be a vessel, a metalworking plant, a school build-ing, a photocopier, or a desk. Some industries, such as oil and gas, have a rigorous process to ensure safe operation and con-struction. This process uses RA methods like Process Hazard Analysis (PHA), Hazard and Operability (HAZOP), What-if Analysis, and Layers of Protection Analysis (LOPA). The con-struction industry also has a thorough design review process. For example, in Singapore, the Building Control Act and its subsidiary regulations provide stringent requirements to ensure that competent designers are engaged to oversee and check designs. However, the Building Control Act is focused on major hazards such as the collapse of a structure and building fires. It does not adequately cover WSH issues like slips, trips and falls, noise-induced deafness, and getting struck by vehicles. DfS Regulations cover all WSH incidents, including non-major incidents involving individual workers' unsafe actions, and have a distinctly wider focus than the Building Control Act.

The concept of DfS is not new to the built environment. For example, the Europeans have been implementing the concept since the 1990s. In the UK, the CDM Regulations have been revised several times and have seen some success. The South African Construction Regulations (2003) and the Australian Occupational Health and Safety Acts have also implemented the concept of DfS. In Singapore, the DfS Regulations were enacted in 2015 and enforced in 2016. DfS is aligned with one of the key principles of the WSH Act: "reducing risks at the source by requiring all stakeholders to eliminate or minimise the risks they create". Despite the differences between the operationalisation of the concept of DfS, all countries implementing DfS seek to minimise the risk of accidents and ill health by considering hazards during the upstream design phases of a construction project.

As seen in Fig. 5.1, as a project progresses from inception to completion, the scope for change decreases and the cost increases. Since DfS involves design changes to eliminate hazards and the implementation of controls to mitigate WSH risks, it becomes more difficult to implement these changes and controls to improve WSH as a project progresses. Therefore, DfS needs to be implemented early in the design phase. However, it is also not possible for DfS reviews to be conducted when there is insufficient information about the design. Thus, DfS reviews must be conducted near the end of each design phase when it is still possible to implement changes. In addition, the DfS review process considers the lifecycle of the structure or product, i.e., it considers the safety and health of workers, occupants, people in the community, maintenance workers, and demolition workers. Therefore, involving construction contractors and facilities managers in the design review process is useful to ensure comprehensive hazard

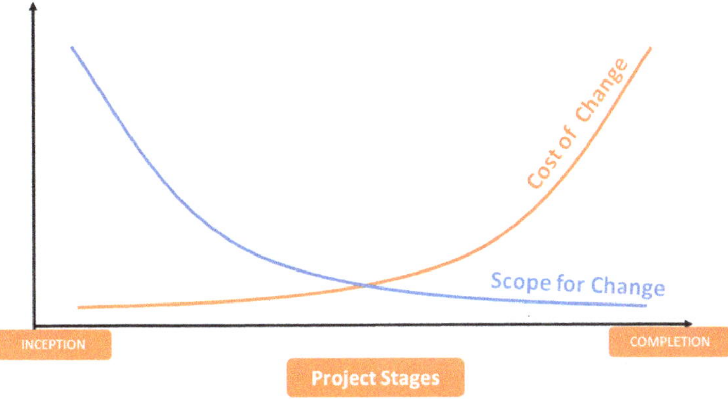

Fig. 5.1 Cost of and scope for design changes to improve safety. (Adapted from (Szymberski,1997)).

identification and suitable risk evaluation. The DfS process will be discussed in more detail subsequently.

5.3 Design for Safety regulations

In Singapore, DfS Regulations apply to projects undertaken by a developer during the developer's business, where the contract sum is S$10 million or more and involves development under Section 3(1) of the Planning Act (Cap. 232). According to the Planning Act, development is "the carrying out of any building, engineering, mining, earthworks or other operations in, on, over or under land, or the making of any material change in the use of any building or land". Section 3(2) of the Planning Act describes specific exclusions not classified as development and are hence not covered by DfS Regulations. For example, the maintenance, improvement, or other alterations of a building that do not materially affect the external appearance or floor area of the building, minor or preliminary works and such temporary use of land as may be declared by the competent authority, and the carrying out by any statutory authority of any works for the purpose of laying, inspecting, repairing or renewing any sewers, mains, pipes, cables, or other apparatus. On the other hand, the Planning Act Section 3(3) highlighted several specific inclusions, e.g.,

- "(a) the use as 2 or more separate houses of any building previously used as a single house involves a material change in the use of the building and of each part thereof which is so used";
- "(b) the use as a dwelling-house of any building not originally constructed for human habitation involves a material change in the use of the building";
- "(c) the use for other purposes of a building or part of a building originally constructed as a dwelling-house involves a material change in the use of the building";
- "(d) the demolition or reconstruction of or addition to a building constitutes development";
- "(e) the use for the display of advertisements of any external part of a building which is not normally used for that purpose involves a material change in the use of the building".

However, it must be noted that the regulations only apply to projects that had a designer appointed on or after 1 August 2016. The DfS Regulations consist of four parts: (1) Preliminary, (2) Duties of Developer, (3) Duties of Designer and Contractor, and (4) Miscellaneous. The following section discusses the duties of the different stakeholders.

5.3.1 Developers

Developers are expected to ensure that "foreseeable risks" are eliminated. If it is not "reasonably practicable" to eliminate the risks, the "design risk" must be "reduced to as low as reasonably practicable" (ALARP) (Chaps. 1 and 4) by reducing the risk at its source and using collective protective measures instead of individual protective measures. In relation to a structure, "design risk" means anything present or absent in the design of the structure that increases the likelihood that an affected person (e.g., construction workers and occupants who work in the structure after construction, including maintenance workers and demolition workers) suffering bodily injury when constructing, working at, or demolishing the structure. The structure refers to any permanent or temporary structures, including the products and mechanical or electrical systems that are part of it. Examples of structures include a school building, a temporary earth retaining structure, a green façade on which a landscape worker works, and a warehouse being demolished.

An example of working at heights is used to illustrate what it means to reduce the risk at source and what collective and individual protective measures are. A DfS review for a condominium building using volumetric construction methods identified that many construction workers need to work at heights to install the volumetric modules during construction. Even though the risk is lower than traditional cast-*in-situ* methods, the installers and lifting personnel will still be exposed to the risk of falling from height. The current approach requires the site workers use fall arrest systems to protect themselves when installing the modules on site. This means the risk is managed downstream (by the contractors) and not at source (by developer and designers). The fall arrest system is an individual protective measure. Even though it is ideal to eliminate the risk of falling from height, it is not feasible with current technology. Thus, following the principle of reducing the risk at source and using collective protective measures, the design team designed a set of foldable guard rails that can be folded when the modules are being transported. The foldable guard rails can then be put into place when the modules are still at the ground level of the site. The worker can use platform ladders or scissors lifts to install the guard rails before they are lifted. The risk is now greatly reduced because the risk control measures (foldable guard rails) are designed for upstream installation and put in place at the prefabrication yard, not at the construction site. The guard rails are a collective protective measure because, in contrast to a fall arrest system, they can protect many workers simultaneously and do not require them to actively ensure that their control measures are in place during work.

Developers must ensure that designers and contractors are competent to perform their duties under the DfS Regulations. The typical approach to ensuring competency is through training. Thus, developers will usually require designers and contractors to have personnel who have taken DfS-related and WSH-related courses. Another way to demonstrate competency is through experience in implementing DfS review processes.

In addition, developers must ensure the project is properly planned and managed. The DfS Regulations specifically high-light the need for "sufficient time and resources" and "relevant information" to be provided to designers and contractors so that they can perform their statutory duties under the WSH Act. However, what constitutes "sufficient time and resources" is not established. One possible approach to determine suffi-cient time and resources is to refer to the time and resources allocated to past similar projects that were safely completed. More research is needed to provide benchmarks to evaluate the sufficiency of time and resources allocated.

Developers need to convene DfS review meetings to iden-tify all foreseeable design risks and discuss how each of the risks can be eliminated or reduced. The DfS processes involved will be discussed in more detail subsequently. Developers must ensure that all relevant designers and contractors attend the DfS review meetings. A DfS register containing information and records of the DfS review meeting and residual design risk must be kept up-to-date and available to designers, contractors, and the regulator (i.e., WSH Inspectors from the MOM). The DfS register is a legal document that must accompany the structure even when the owner changes. The developer must inform any new owner of the nature and purpose of the DfS register.

The developer can delegate the duties related to the DfS review meeting and DfS register to a "DfS Professional"

(DfSP) — a person assessed by the developer to be competent to perform the duties. The WSH Council approved some industry associations, such as the Institution of Engineers Singapore (IES) and the Singapore Contractors Association Ltd (SCAL), to conduct a two-day DfSP course to certify DfSPs. Participants are required to have significant experience in the construction industry. The course also requires participants to submit a report containing a DfS register for assessment.

The developer must provide all the necessary information for the DfSP to perform the delegated duties. On the other hand, the DfSP must, as soon as reasonably practicable after each DfS review meeting, provide the developer with all relevant information on each foreseeable design risk identified at the meeting and how each design risk can be eliminated or reduced. In addition, the DfSP must, as soon as reasonably practicable after any information or record is added to the DfS register, provide the developer with an updated copy of the DfS register. It must be noted that the developer cannot delegate its general duties under the DfS Regulations, which are ensuring competent designers and contractors are engaged, ensuring that the project is properly planned and managed with sufficient time and resources for designers and contractors, and ensuring designers and contractors have all relevant information.

From the duties allocated in the DfS Regulations, it is obvious that developers have an important role to play in DfS. This is a suitable approach because past research (Goh and Chua, 2016; Toh *et al.*, 2016) has shown that developers can motivate designers to implement DfS.

5.3.2 Designers

In the DfS Regulations, a "designer" is the person who prepares a design plan relating to a structure. "Design plans" include drawings, building information modelling, design details, specifications, materials and bills of quantities (including specifications of articles or substances) relating to a structure, and calculations prepared for a design. This definition is very broad and can even include quantity surveyors who develop bills of quantities that influence the safety of the design.

While preparing the design plan, designers (particularly engineers, architects and contractors) must "as far as reasonably practicable, eliminate all foreseeable design risks". Suppose it is not possible to eliminate the design risk. In that case, the designer must propose to the person who appointed the designer, which can include the developer, designer or contractor, a modification to the design plan that reduces the design risk to ALARP. As discussed earlier, the modification must consider the principle of reducing risk at source and adopt collective protective measures instead of individual protective measures. Designers must provide all relevant information on the structure's design, construction or maintenance to the person who appointed the designer.

Many designers do not have sufficient experience and knowledge of the downstream processes, including construction, maintenance, and demolition works. The early involvement of construction contractors and facilities managers can minimise this knowledge gap. Concurrently, training designers on WSH hazards and controls are useful to better assess WSH issues during the design phase.

5.3.3 Contractors

Contractors must inform the person who appointed them of any foreseeable design risk that they know of. They must also ensure the designers and subcontractors engaged by them are competent and provide them with relevant information.

Contractors are heavily regulated in terms of site WSH management and have the most direct control over site safety. In contrast, DfS is more focused on developers and designers. As discussed earlier, it is advantageous for contractors to be involved in the design as early as possible so that they can provide inputs on the design before it is frozen. Contractors also engage designers to design temporary structures. In this context, they must also ensure that the designers are versatile with the DfS process.

In the case of design and build projects, the contractor performs the role of both a designer and a contractor. The contractor still needs to engage architects and engineers to conduct the design in actual implementation. Since the contractor engages these designers, the project has the advantage of closer collaboration between the contractor and designers.

5.4 Risk assessment and Design for Safety review

Chapter 4 discussed risk management and risk assessment (RA) focused on operations and work processes. Since DfS is a design risk assessment, it is similar to a contractor's risk assessment or operational RA, but there are differences. Table 5.1 summarises the key differences.

Note that a contractor's risk assessment and DfS review overlap and are not mutually exclusive. They should complement each other. The DfS review process aims to form the foundation for the contractor's RA, i.e., the contractor's RA should use the DfS reviews as inputs. It is also important to note that the DfS review should not turn into an operational risk assessment that contractors do. The DfS review should focus on what the designers can do to improve WSH. For example, the contractor's RA would be more concerned about workers' behavioural issues, PPE, and administrative controls. On the other hand, the DfS review is more concerned with using design changes to eliminate and mitigate design risks. The key is to focus on high-risk hazards whenever there are overlaps between the contractor's RA and DfS review. Designers should always seek ways to reduce the risk of high-risk hazards that

Table 5.1 Differences between a contractor's risk assessment and DfS review.

Contractor's risk assessment	DfS review
Hazards identified based on work activities, resources used, and locations	Hazards identified based on design elements and considerations
Controls are typically focused on lower rungs of the Hierarchy of Control: engineering control, administrative control, personal protective equipment (PPE), and Safe Work procedures (SWP)	Controls are typically focused on higher rungs of the Hierarchy of Control: elimination, substitution, and engineering control
Address the operations during the construction stage	Address the full life cycle of the structure
Focus on what contractors can do to improve WSH	Focus on what the designers can do to improve WSH during the design process

are hard for contractors to manage. For example, falling from height is the most common cause of fatalities in the construction industry, so designers should always try to eliminate or reduce the need to work at height during construction and maintenance through better designs.

5.5 Design for Safety review process and tools

This section discusses a DfS review flowchart developed by the author and other available approaches to facilitate the DfS review process, e.g., CHAIR (Construction Hazard Assessment and Implication Review) and GUIDE.

5.5.1 Design for Safety review flowchart

Figure 5.2 provides a flowchart to guide a DfS review process. The first step of a DfS review is to identify hazards across the project

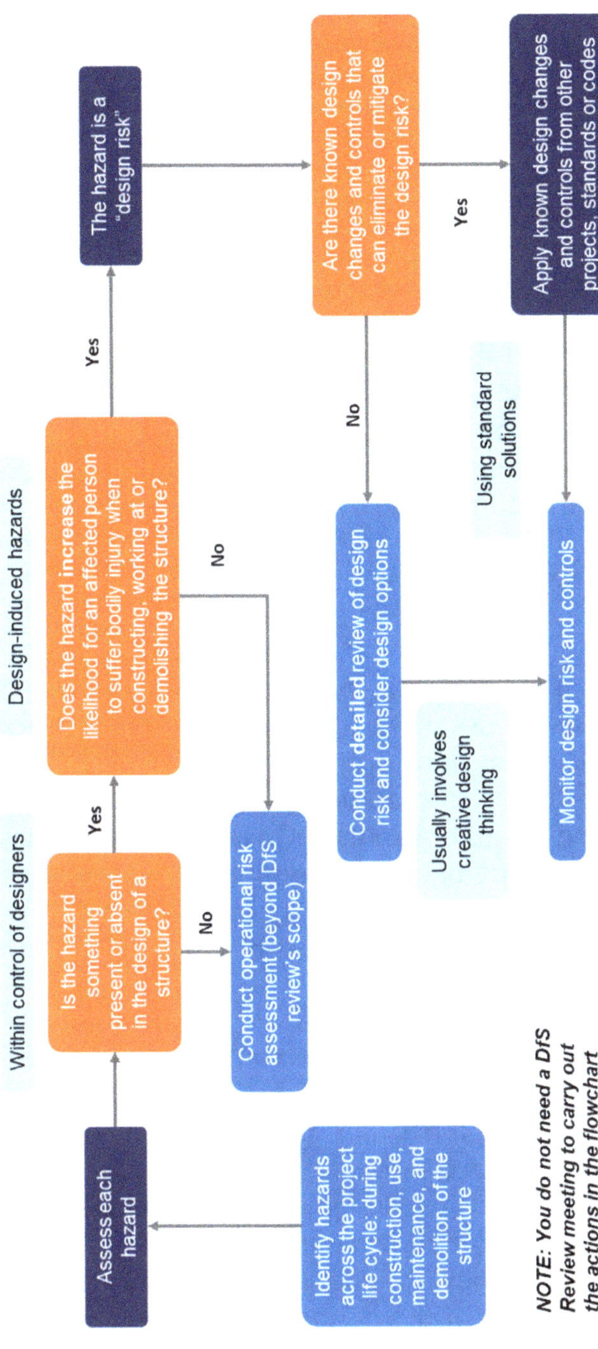

Fig. 5.2   DfS review flowchart.

lifecycle. These hazards should be identified as early as possible so that designers can eliminate or reduce the design risks. Each hazard that is identified should be assessed. One question to identify hazards within designers' control is, "Is the hazard something present or absent in the design of a structure?" If a hazard is unrelated to the design, it is not within the designers' control. The DfS review team can park the hazard aside and let the contractors do their job doing their operational risk assessment.

Suppose the hazard is related to the design of a structure. In that case, the next step is to evaluate whether the hazard increases the likelihood of an affected person suffering bodily injury during work. If the hazard does not increase the likelihood of an injury, it need not be covered in the DfS review. If the two conditions of (1) hazard being related to a structure's design and (2) increasing the likelihood of injury are met, the hazard is a design risk.

Once identified as a design risk, the DfS review team should aim to reduce the design risk to ALARP. The team should check for known design changes and controls that can eliminate or mitigate the design risk. If these known design controls and changes are accepted norms or stipulated in industry guidelines and standards, they are usually considered reasonably practicable. In such situations, designers are expected to implement them, but they should still consider possible exceptions that might render the controls ineffective.

Suppose no known design changes and controls can eliminate or mitigate the design risk. In that case, the DfS review team must thoroughly review the design risk and consider different design options. It will usually involve creative design thinking, where the DfS review team understands the design risks and thinks of ways to design them out. A detailed review is more resource-intensive, and the DfS review team should consider a few factors. Firstly, the risk level of the design risk. If the design is of medium or high risk, especially if the design is unique, innovative, or new, with no applicable code or standards, then a detailed review is suitable. Secondly, the DfS review team should also consider if there are difficulties controlling the design risk downstream. Past experiences might show that specific hazards are especially hard to control downstream. For example, there are many fatalities due to falling from heights and being struck by moving plants. Therefore, designers need to think of ways to reduce these design risks as far as reasonably practicable.

On the other hand, if the design risk is low, there are effective downstream controls, and the identified design changes and controls are not reasonably practicable. The team might not need to implement any controls or changes. However, the DfS review team must demonstrate that they have thought through the issues and can justify their decision. For example, they should have considered past statistics and accident cases and reviewed the effectiveness of controls implemented by downstream stakeholders.

Suppose the DfS review team decides not to implement any controls or design changes for an identified design risk. In that case, they should inform the contractors of the design risk so that they can manage it more effectively.

Regardless of whether the controls are from the detailed review or known solutions, it is important to continue monitoring the design risk and controls. Monitoring is important because no one can be sure that the assessments are accurate and the solutions are effective when implemented. Thus, the DfS review team should monitor the design risks to prevent them from posing a hazard to workers.

It should be noted that designers should conduct the actions indicated in the flowchart (Fig. 5.2) on an ongoing basis and not wait for a DfS review meeting. This is because designers make decisions on a day-to-day basis that impact WSH. Hence, they must keep DfS in mind throughout the design process.

Fig. 5.3 Example from the IES-NUS DfS Library (2021).

We will discuss a DfS review example (Fig. 5.3) from the IES-NUS DfS Library (2021). There is a worker working on a link bridge. The worker is part of a construction team trying to construct the roof link bridge. The structure requires construction and maintenance workers to work at height during different stages of the lifecycle. Is this hazard a design risk that requires design changes or controls?

In this case, the design consideration is the link bridge between buildings. The broader context is the link bridge with the roof requires workers to work at height during construction and maintenance. The design risk is the link bridge with open edges. Since the link bridge has open edges, it increases the likelihood of workers falling from height. With reference to the flow chart in Fig. 5.2, the activity is during the construction and maintenance stage, and the hazard is the open edges, which can lead to falling from height because workers need to work at heights and there are open edges. The link bridge roof and the open edges are part of the structure, so the hazard is related to the design. Since accident statistics show that a significant percentage of construction fatalities are due to falls from height, it is important for designers to try to reduce this design risk.

The next question is, "Are there known changes or controls that can mitigate the design risk?" Even though there are downstream controls such as edge protection and barricades, designers should be more involved and consider design alternatives that can reduce the risk through collective protective measures. For example, they can design for the assembly of the link bridge roof to be conducted on the ground and then lift it into place. This design change would require a detailed review of the design risk and consideration of different design options.

If we refer to the definitions, hazard means anything with the potential to cause bodily injury. Design risk is anything present or absent in the design of the structure that increases the likelihood that an affected person will suffer bodily injury. Furthermore, operational hazards, such as construction and maintenance hazards, mean anything present or absent during downstream activities. These definitions overlap, as we have seen in this example. If the designers do not implement any design changes or controls, the contractor would still put in controls to manage the design risk, which is also an operational hazard that contractors manage. Thus, there is always a grey area between the two types of hazards. In the example given, the grey area can cause confusion as to whether the hazard should be handled as part of the DfS review or during operational risk assessment. DfS review teams need to exercise their judgement through a structured DfS review process (Fig. 5.2) to ascertain if the hazard is a design risk that should be evaluated and controlled during the DfS review.

The possible controls can be derived based on the hierarchy of controls (Chap. 4), where elimination, substitution, and engineering control are part of the safe design approach. For example, if the link bridge is assembled on the ground and lifted into place, it eliminates the need for workers to be on the link bridge roof at heights. Another option is to install the bridge at a lower level as a substitution, which may not be feasible. In terms of engineering control, designers can design convenient ways for the guardrails to be installed earlier, to reduce construction and maintenance workers' exposure to the open edge. From the contractor's point of view, they frequently rely on PPE, such as a fall arrest system, and administrative controls such as work-at-height training and a permit-to-work system to reduce the risk of fall from the link bridge. These controls are typically less effective. Therefore, it is important for designers to use the DfS review to implement safe design approaches to reduce the risk to workers.

In the UK, instead of the hierarchy of control, the acronym ERIC is used to guide designers in selecting ways to manage design risks. ERIC stands for Eliminate, Reduce, Inform, and Control. Eliminate is the same as elimination in the hierarchy of control. Reduce and Control are about the use of the design process to reduce or control design risks when it is not reasonably practicable to eliminate them. One way to differentiate Reduce and Control is to relate Reduce to substitution in the hierarchy of control and Control to engineering control. Inform is about passing details of design risks to other stakeholders, i.e., developer, other designers, and contractors.

5.5.2 Design for Safety guidelines by Workplace Safety and Health Council, Singapore

The DfS Guidelines (Workplace Safety and Health Council, 2022) describe a recommended DfS review and documentation process. The overall DfS review process is based on the GUIDE acronym, which represents:

1. **G**roup together a DfS review team consisting of main stakeholders.

2. Understand the design concept by looking at the drawings and calculations or have designers elaborate on the design.
3. Identify the hazards and risks of the design or construction method. The risks and hazards should be recorded and analysed to see if they can be eliminated by changing the design.
4. Design around the hazards and risks identified to eliminate or mitigate the risk.
5. Enter all information into the DfS register, including residual risks to be mitigated and information on vital design changes that would affect the safety and health of affected persons.

According to the 2016 version of the DfS Guidelines (Workplace Safety and Health Council, 2016), the GUIDE process should be conducted at three junctures, each with a different focus. The three design reviews are named GUIDE-1, GUIDE-2 and GUIDE-3. For a traditional design-then-build contract, a GUIDE-1 design review is conducted after the concept design is nearly completed, but the design still has room for amendments. It is meant to focus on the project's general location, traffic, type of buildings in the surroundings, and other general characteristics. If necessary, more than one meeting should be conducted at each stage to review all the possible hazards arising from the design.

GUIDE-2 is focused on the detailed design, maintenance and repair. GUIDE-2 should review the hazards related to construction methods, access and egress, and whether the design will create confined spaces or other hazards. Hazards during the maintenance and repair of the structure should also be reviewed. Even though GUIDE-2 should include information provided by the contractor, this will be challenging for a traditional design-then-build project as the contractor may not be appointed yet. However, in alignment with approaches such as early contractor involvement (ECI), contractors can be engaged as consultants to improve the DfS review process. Since hazards during the maintenance and repair phase will be identified, eliminated or mitigated, it is also useful to involve experienced facilities managers and maintenance contractors during GUIDE-2.

A design-and-build (D&B) contract has the advantage of integrating the contractor and lead designers into one. In the case of a D&B contract, the contractor can provide construction method information earlier and influence the design to eliminate or mitigate risk at the construction stage.

GUIDE-3 is focused on the pre-construction review, particularly on temporary works design and design by specialist contractors not covered during the concept and detailed design phases. Some key hazards and controls include shoring, trenches and deep excavation, confined spaces, formwork, and falsework. Residual risk of hazards and controls will have to be documented and highlighted to the contractors at the end of GUIDE-3. For example, suppose foldable guard rails are designed to protect workers working at the top of prefabricated modules during installation. In that case, the folded guard rails need to be put into place by site workers before lifting the modules. In this case, the residual hazard is falls from height when workers set up the guard rails. The contractor will have to be notified of this hazard, assess the risk before and during construction, and put control measures in place to ensure that the risk is ALARP.

The DfS Guidelines highlight the following aspects that must be covered (but not limited to) during the GUIDE process:

• General design concept	• Layout
• Accessibility	• Maintenance
• Confined space	• Material handling or storage
• Emergency	• Means or methods
• Lighting	• Operation
• Excavation	• Physical hazards
• Fall prevention	• Sequence of construction
• Working platforms	• Standardisation of building elements
• Hoisting or weight	• Weather

As with any RA, the DfS review process must be structured and easy to administer. To ensure efficiency and effectiveness during the DfS review sessions, it is important to predetermine a list of design considerations that should be discussed during the design review meeting. These design considerations can be identified through a preliminary evaluation conducted before the design review meeting through the use of checklists, guidewords (see the CHAIR method discussed later), and what-if questions on possible design-related hazards. The developer or his representative (e.g., a DfSP) will then compile the returns from all stakeholders. For example, the design considerations during a GUIDE-2 review can include air-con ledges, a skylight in the atrium area, a building maintenance unit, etc. These considerations can lead to identification of hazards and design risks such as open edges, lightning, and falling objects. The compiled list of design considerations and design risks can then be used during DfS review meetings to facilitate the review process.

It must be noted that the review should focus on design risks and controls not already covered in the Building Control Act and the Workplace Safety and Health Act. Since designers and contractors are already expected to comply with relevant regulations, the DfS review meetings should dedicate time to identifying design risks peculiar to the constructed structure. For example, if the construction site is next to a hospital, traffic to the hospital can be affected if the site entrance is placed along the road leading to the hospital. This can delay emergency vehicles trying to reach the hospital. In this case, even if having the entrance along the road to the hospital does not contravene any regulations, the entrance should be shifted. Compliance checks must still be conducted, but individual designers should be capable of doing that within existing building control processes. Nevertheless, developers and designers must be sensitive to WSH information, changes and incidents that may suggest a need to review specific designs even if they are deemed "standard" designs.

Another area of focus during DfS reviews is the interface between design considerations and construction, maintenance, and operational activities. DfS review meetings should consider the activities occurring during the relevant activities related to relevant design considerations and the possible hazards that can arise. For example, during precast construction, where key design considerations include the different types of precast components, the delivery of precast components will mean the presence of trailers and increased lifting operations. These activities will increase the risk of being struck by moving vehicles and being hit by objects falling from height. This means that during GUIDE-2 or GUIDE-3, the site layout design should consider the shape, length and size of large precast components and ensure that site workers and delivery and lifting activities are segregated. These design changes should be tied to the different design considerations.

Even though the GUIDE method is useful, the 2022 version of the DfS Guidelines (Workplace Safety and Health Council, 2022) indicated other DfS tools can also support the DfS review process. For example, Fig. 5.4 shows the possible tools that can be used during the different stages of a design and build project. Some of these tools will be reviewed in the subsequent sections.

The appendix in this chapter shows a typical design risk assessment form to record the DfS review process (Workplace Safety and Health Council, 2022). The design risk assessment form should be updated frequently as new information becomes available and actions are completed. Hence, the form should ideally be stored in digital format. The DfS review team must append other supporting information, e.g., drawings, material specifications, and equipment manual, with the design risk assessment form, which forms the DfS register. The DfS register should also contain the key outcomes or decisions made regarding the design. It could include the selected design changes and controls described using details such as shape, location, material, and dimensions of design elements. Whoever owns the structure must have a copy of the DfS register.

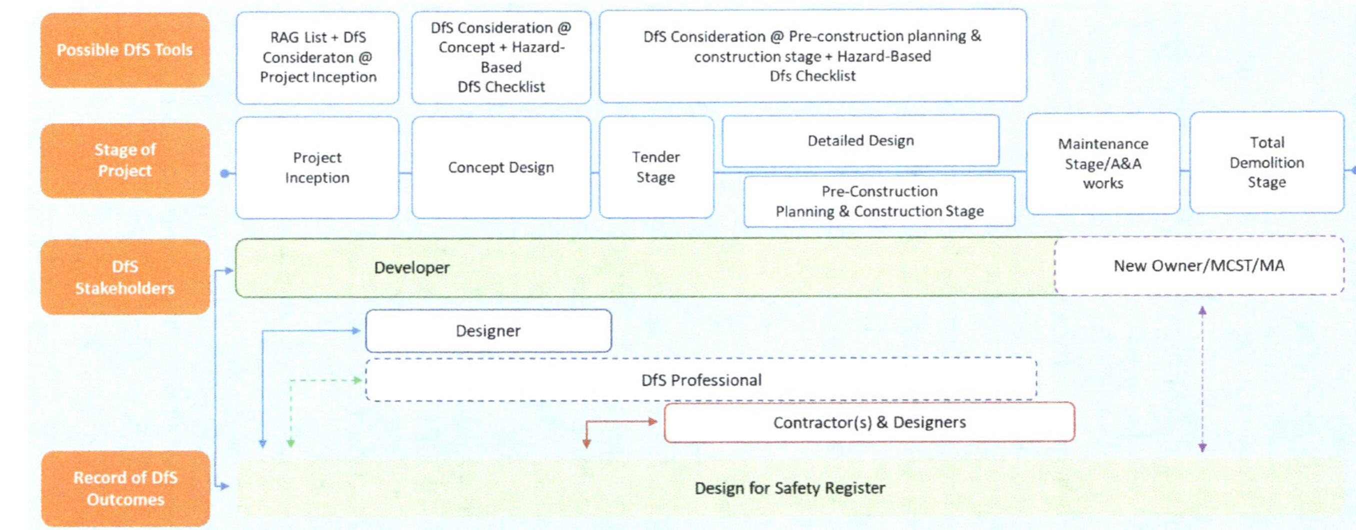

Fig. 5.4   Sample DfS process for a Design & Build project (New Build) (Workplace Safety and Health Council, 2022).

5.5.3 CHAIR: Construction Hazard Assessment and Implication Review

The Construction Hazard Assessment and Implication Review (CHAIR) was created by WorkCover New South Wales (2001). CHAIR is essentially a customisation of the HAZOP method (see Chap. 4 for more details) for the construction industry. CHAIR aims to identify and eliminate hazards or minimise risk levels of the hazards in a design as early as possible. The process involves all key stakeholders, who gather to reflect on the design from a health and safety perspective. CHAIR is implemented at two key junctures in the project lifecycle: during the conceptual design stage (CHAIR-1) and just before construction (CHAIR-2 and CHAIR-3). CHAIR-1 is meant to review the concept design, CHAIR-2 is meant to review construction and demolition issues, and CHAIR-3 focuses on maintenance and repair issues.

CHAIR involves the following steps:

1. Assembling a CHAIR study team (including all stakeholders).
2. Defining the objectives and the scope of the study.
3. Agreeing on guidewords or prompts to assist in the brainstorming process.
4. Partitioning the design (CHAIR-1, CHAIR-3) or construction process (CHAIR-2) into logical blocks of appropriate size.
5. For each logical block, using various guide words to assist with identifying safety aspects/issues.
6. Discussing associated risks and determine if the safety risk can be eliminated.
7. If the safety risk cannot be eliminated, determining how it might be reduced.
8. Assessing whether the proposed risk controls (e.g., expected safeguards, etc.) are appropriate (i.e., is the risk ALARP?).
9. Documenting comments, actions and recommendations — determining appropriate solutions for design issues still to be resolved.

Since HAZOP is known for its rigour in assessing the safety and operability of a design, typically for process plants, CHAIR, being a customised form of HAZOP, is also rigorous. However, HAZOP is not well-used in the construction industry due to difficulties in creating guidewords and the time to conduct a HAZOP study. A brief enquiry with a major Australian construction contractor and an experienced Australian occupational safety and health expert in the construction industry indicates that CHAIR is apparently not well-used in Australia. In addition, Bluff (2003) discussed that the Memorandum of Understanding (MOU) signed by 17 contractors in New South Wales (NSW), Australia, which provided support for the development of CHAIR, did not have a significant impact on DfS in NSW. Bluff (2003) indicated that the challenges in implementing DfS are: too much focus on paperwork; failure to address safety in design by clients, design profession, and principal; poor programming practices; and unrealistic timeframes. It can be hypothesised that the usefulness of CHAIR as a DfS tool is highly dependent on the safety attitude of clients, designers and contractors, adequate planning, and the provision of a reasonable timeframe. This is related to the concept of DfS climate, which is discussed in Chap. 7.

5.5.4 Safe design process in Australia

The Safe Design (SD) process in Australia is similar to the GUIDE process in Singapore, but it provides further details useful for DfS reviews. The SD process is described in the flowchart in Fig. 5.5, and a summary of the process, as described in Safe Work Australia (2018), is provided below.

5.5.4.1 Pre-design stage

According to Safe Work Australia (2018), the pre-design phase of the SD process includes:

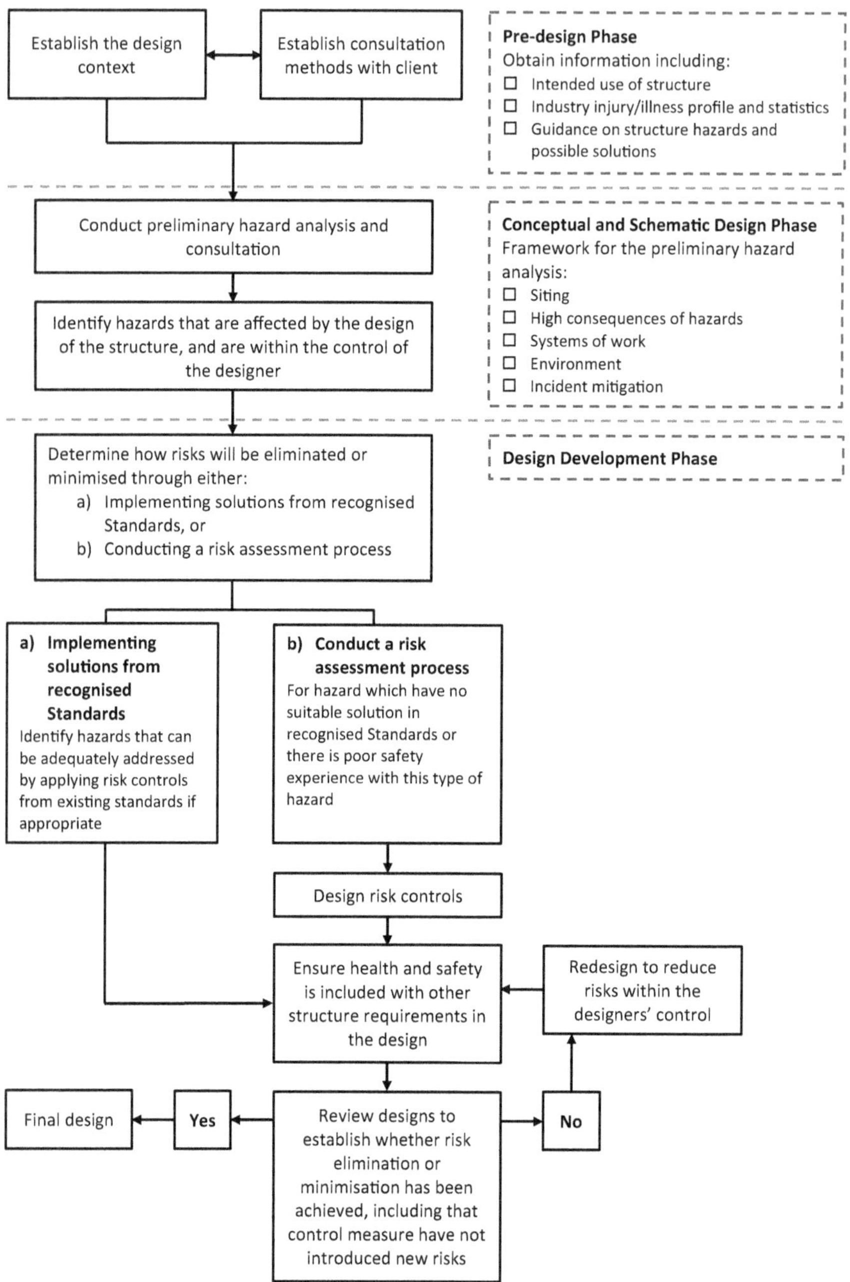

Fig. 5.5 Safe design systematic approach. (Adapted from (Safe Work Australia, 2018, Fig. 1)).

- Establishing the design context, focusing on the purpose of the structure, as well as the scope and complexity of the project.
- Establishing the risk management context by identifying the breadth of workplace hazards and relevant legislation and codes of practice and standards that need to be accounted for.
- Identifying the required design disciplines, skills and competencies.
- Identifying the roles and responsibilities of various parties in the project and establishing collaborative relationships with clients and others who influence the design outcome.
- Conducting consultations and research to aid in identifying hazards and assessing and controlling risks.

In this stage, the role of the client is to prepare a project brief that includes the safety requirements and objectives for the project, ensuring a shared and comprehensive understanding of safety expectations between the client and designer. Additionally, the client must provide the designer with all available information relating to the site that may affect safety and health. The designers, on the other hand, are required to ask their clients about the type of activities and tasks likely or intended to be carried out in the structure, including the tasks of those who maintain, repair, service, or clean the structure as an important part of its use. Information required for the pre-design stage can be obtained from sources such as WSH and building laws, technical standards and codes of practice, industry statistics regarding injuries and incidents, hazard alerts or other reports from relevant statutory authorities, unions and employer associations, specialists, professional bodies representing designers and engineers, and research and testing done on similar designs.

5.5.4.2 Conceptual and schematic design stage

In this stage, preliminary hazard analysis should be conducted as early as possible. Firstly, the designers and other relevant stakeholders or duty holders involved in the preliminary hazard analysis should decide which hazards are "in scope". A hazard is "in scope" if "it can be affected, introduced or increased by the design of the structure". At this stage, preliminary considerations should be given to possible ways to eliminate or substitute hazards.

Next, systems of work that are foreseeable as part of the construction method and the intended use of a structure as a workplace should also be identified in the preliminary hazard analysis. Likely or intended workflows, if known, will also be useful as part of the project brief prepared by the client. Furthermore, the project brief may also include any activities and systems with hazards specific to the nature of the structure. In the case of hazardous manual tasks, Australian WSH Regulations (WHS Regulations 61) specify that a designer must, as far as reasonably practicable, ensure that a structure is designed to eliminate the need to carry out any hazardous manual task in connection with it. Where this is not reasonably practicable, the designer must ensure the structure is designed so that the need for these hazardous manual tasks is minimised so far as is reasonably practicable.

Moreover, a systematic approach towards hazard identification should be done. The recommended framework in the Australian Code of Practice is elaborated in Table 5.2. It is noted that, as in the case of the Singapore DfS Guidelines, the use of checklists or hazard prompt lists is a useful preliminary hazard identification step. Still, these lists must not limit SD or DfS review teams, and they should explore hazards beyond the lists.

5.5.4.3 Design development phase

During the design development phase, the design team converts the design concepts into detailed drawings and technical specifications. The team will decide on design-related control measures and prepare construction documentation. The completed design is then handed over to the client.

Table 5.2 Framework for preliminary hazard identification. (Adapted from (Safe Work Australia, 2018, Table 2 of Code of Practice for Safe Design of Structures)).

Source of hazard	Description
Siting of structure	Potential design issues that may affect safety include: • proximity to adjacent property or nearby roads • surrounding land use • clearances required for construction equipment and techniques • demolition of existing assets • proximity to underground or overhead services — especially electricity lines • exposure of workers to adjacent traffic or other hazards • site conditions — including foundations and construction over other assets or over water • safety of the public • use of adjacent streets
High consequence hazards	• Storage and handling of dangerous goods • work with high energy hazards (e.g., pressure) • health hazards such as biological materials
Systems of work (involving the interaction of persons with the structure)	The systems of work (including cleaning and maintenance activities) that pose risks, for example: • rapid construction techniques, i.e., prefabrication versus *in-situ* construction • materials to be used in construction • staging and coordination with other works • inadequate pedestrian or vehicle separation • restricted access for building and plant maintenance • hazardous manual tasks • working at height • exposure to occupational violence Consider both technical and human factors, including humans' ability to change behaviour to compensate for design changes. Anticipate misuse throughout the lifecycle.

(Continued)

Table 5.2   (*Continued*)

Source of hazard	Description
Environmental conditions	• Impact of adverse natural events such as cyclones, floods and earthquakes • inadequate ventilation or lighting • high background noise levels • facilities that do not meet workplace needs
Incident mitigation	A structure can increase the consequences of an incident if it has: • inadequate entry and exit points • poorly situated assembly areas • inadequate access for emergency services

The design development phase should involve:

• Developing a set of design options corresponding to the hierarchy of control.
• Selecting the optimum solution. Balance the direct and indirect costs of implementing the design against the benefits derived.
• Testing, trialling or evaluating the design solution.
• Redesigning to control any residual risks.
• Finalising the design and preparing the safety report and other risk control information needed for the structure's lifecycle.

Legislative provisions, guidelines and standards must be followed when implementing solutions or control measures. However, the mandatory regulatory requirements and standards may not adequately control WSH risks if applied to unique situations or if the requirement or standard is outdated. Therefore, the DfS review process must account for situations not covered by regulatory requirements and standards.

Moreover, RAs should be done to determine the severity of the risk, the effectiveness of existing control measures, the actions one should take to control the risks, and the urgency of such actions to be taken. Risks are assessed by considering what could happen if someone is exposed to a hazard and the likelihood of it happening. In cases where the hazards and their associated risks are well-known, well-established and accepted, effective control measures are used in a particular industry suited to the workplace circumstances. A formal RA is not required, and the controls can be implemented immediately. However, this author recommends that there must be effective monitoring processes to ensure that the control measures continue to be effective throughout its implementation. The Australian Code of Practice recommended several RA methods for assessing design safety:

- fact-finding to determine existing controls, if any;
- testing design assumptions to ensure that aspects of the design are not based on incorrect beliefs or anticipations on the part of the designer, for example, as to how workers or others involved will act or react;
- testing structures or components specified for use in the construction, end-use and maintenance phases;
- consulting with key people who have the specialised knowledge and capacity to control or influence the design (for example, the architect, client, construction manager, engineers, project managers, and safety and health representatives); and consulting directly with other experts (for example specialist engineers, manufacturers, and product or systems designers) who have been involved with similar constructions;
- when designing for the renovation or demolition of existing buildings, reviewing previous design documentation or information recorded about the design structure and any modifications undertaken to address safety concerns, and consulting professional industry or employee associations who may assist with RAs for the type of work and workplace.

Table 5.3 provides a framework to ensure all risks are addressed in the design and identifies the parties that should be involved in the various steps.

Table 5.3 Design process. (Adapted from (Safe Work Australia, 2018, Table 3 of Code of Practice for Safe Design of Structures)).

Step	Possible techniques	By whom
Identify solutions from regulations, codes of practice, and recognised standards	• Consult with all relevant persons to determine which hazards can be addressed with recognised standards. • Plan the risk management process for other hazards.	Designer led. Client approves decisions.

Apply risk management techniques	• Further detailed information may be required on hazards, for example, by:
	   ○ using checklists and referring to codes of practice and guidance materials, and
	   ○ job/task analysis techniques.
	• A variety of quantified and/or qualitative RA measures can be used to check the effectiveness of control measures.
	• Scale models and consultation with experienced industry personnel may be necessary to achieve innovative solutions to longstanding issues that have caused safety problems.
		Designer led. Client provides further information as agreed in the planned risk management process.

Discuss design options	• Take into account how design decisions influence risks when discussing control options.	Designer led. Client contributing.
Design finalisation	• Check that the evaluation of risk control measures in the design is complete and accurate. • Prepare information about risks to health and safety for the structure that remains after the design process.	Designer led. Client and designer agree on the final result.

Table 5.3   (*Continued*)

Step	Possible techniques	By whom
Potential changes in construction stage	• Ensure that changes that affect design do not increase risks; for example, the substitution of flooring materials could increase slip/fall potential and may introduce risks in cleaning work.	Construction team in consultation with designer and client.

5.5.4.4 Reviewing control measures

As the design process continues, more detailed decisions are made. These decision points become opportunities for either eliminating or minimising risks. Designers should deliberately review design solutions at each decision point to establish the effectiveness of the risk controls previously selected and change them if necessary and so far as is reasonably practicable.

As indicated earlier, it is best if SD or DfS reviews involve the contractor and facilities management personnel with knowledge and experience in construction and maintenance processes. Their expertise will assist in identifying WSH issues that are overlooked during design.

The effectiveness of the control measures arising from the DfS review should be evaluated after construction. This ensures the identification of the most effective design practices and any design innovations that could be applied to future projects. The review can be conducted in a post-construction workshop attended by all relevant parties. On top of that, subsequent feedback from users to assist designers in improving their future designs may be provided through post-occupancy evaluations, defects reports, accident investigation reports, information regarding modification, user difficulties, and deviations from intended conditions of use.

5.5.5 Online libraries supporting Design for Safety reviews

Some online libraries that can support DfS reviews include:

- Institution of Engineers, Singapore (IES) (2018), https://www.ies.org.sg/Publication/Technical-Resources
  - Design for Safety (DfS) Library: Examples of Hazards–Architectural Design
  - Design for Safety (DfS) Library: Examples of Hazards–Mechanical & Electrical Design
  - IES-NUS Design for Safety (DfS) Library for Designers: Construction and Maintenance Design Risks
- Real Estate Developers' Association of Singapore (REDAS) (2019), https://redas.com/publication/dfs-good-practice-guide/
  - DfS and WSH good practice guide
- UK Design Best Practice — Promoting safety in design, http://www.dbp.org.uk/
  - Topics include Architecture, Civil, Structural, Mechanical, Electrical, Buildability, and Health & Wellbeing
- The Occupational Safety and Health Council (2016) developed a series of worked DfS review examples in Hong Kong, https:// www.devb.gov.hk/

These online libraries are useful, but the IES and Hong Kong resources are rather static and can be more comprehensive. The UK DBP database is relatively comprehensive, but more can be done to improve the search function and standardisation of the information in the database. More can be done to improve these resources.

5.5.6 Other Design for Safety tools

The range of DfS tools can be very wide — from rule-based Building Information Modelling (BIM) software developed to highlight possible design risks (e.g., (Hossain *et al.*, 2018)) to simple checklists. Existing BIM clash detection tools can also help designers identify clashes between building elements, but such functions are based on pre-defined rules to identify WSH hazards automatically. Defining and maintaining these rules can be a challenge.

The Red-Amber-Green list is a DfS tool used in the UK. It is essentially a list of design risks that are pre-defined to be high-risk (red), medium-risk (amber), and good practices (green). For example, some of the red-list design risks that designers might want to avoid include fragile roof materials, slippery tiles, and materials with poor fire safety ratings. Amber-list design risks are lower risk but have significant residual risks that should also be avoided when possible. Some examples include external manholes near high-traffic areas, large and heavy glass panels, and site layouts with two-way traffic. Green-list items are encouraged, for example, giving adequate workspace and headroom for maintenance in the plant room, windows that can be cleaned from inside, and early provision of edge protection.

Another tool is guide word lists. These guide words are meant to help the DfS review team brainstorm possible design risks, which is similar to the CHAIR method discussed earlier — for example, heights, high energy, pressure, and temperature. Reviewing these guide words helps the team to try to imagine possible hazards.

There are also DfS checklists. The DfS Guidelines have a comprehensive checklist containing questions such as: "Can less hazardous materials be used (e.g., solvent-free or low solvent adhesives and water-based paints)?" and "Can materials that create significant fire risk be removed?" These checklist questions help the DfS review team to identify possible design risks.

5.6 Types of design risks

Based on the UK Collaborative Reporting for Safer Structures (CROSS-UK) (2021), there are three main types of design risks.

Type A Design Risks are typical workplace safety and health hazards. They are related to work activities covered by an operational risk assessment. Some examples include work-at-height, use of site vehicles, electrical/gas works, trench work, noise-related issues, vibration, and dust. Contractors typically manage these, but these hazards are also design risks. If the hazard is a design risk, designers may still need to develop design changes and controls to reduce the risk during construction.

Type B Design Risks are codified design risks that are covered in design codes. They tend to focus on the prevention of collapse. For example, structural collapse, overturning, fire resistance of materials, fire spread, and means of escape are generally well covered by codes. Nevertheless, designers need to keep an eye for exceptions to the codes.

Type C Design Risks are non-codified hazards and risks. They are the "soft" hazards originating from the design. Even though they lead to a risk to others, they do not lead to specific work tasks on-site or on the completed structure. Some examples are calculations, design concept errors, analysis errors, or lack of adequate supervision. These design risks can lead to major collapses and can be very dangerous. In many instances, they are not well-managed and are the common cause of collapses. The problem with this type of hazard is that they are not directly overseen by the contractors, and the contractors may not be able to spot these hazards easily until there are clear signs of problems. However, their workers are directly affected by these hazards. Thus, it is important for designers and contractors to be fully aware of such soft hazards.

As identified in the CROSS-UK (2021) safety alert, Type C hazards are typically due to people and processes; they are similar to the underlying factors in the ECT. In terms of people, the major concerns are competency, time, and resource issues. These issues are related to DfS because developers are expected to employ competent designers and contractors and provide sufficient time and resources. Thus, a well-implemented DfS regime can reduce Type C design risks. In terms of processes, there are many possible weaknesses, such as unsafe design assumptions, lack of proper constructability review, poor communication and follow through on design checks, lack of clarity of design responsibilities, no single point responsibility, poor change control, and lack of independent supervision. These Type C hazards can be reduced with a thorough DfS review process. If an organisation can prevent Type C hazards, it can prevent Type B hazards (codified hazards) because the thorough review process will ensure that relevant codes are applied correctly. In addition, since a key purpose of DfS review is to reduce Type A downstream hazards during construction, use, maintenance and demolition of the structure, a well-conducted DfS review process will also reduce Type A hazards. Thus, DfS can reduce the risk of all three types of hazards.

5.7 Common misconceptions

There are six common misconceptions about DfS. Firstly, many people think that it is compulsory to engage a DfSP for all projects with a contract sum greater than S\$10 million. They think they need a DfSP before they can conduct the first DfS review. In fact, the DfS Regulations clearly state that the DfSP is not compulsory. It is an option for developers to delegate some of their duties to a DfSP. Even if they appoint a DfSP, developers still retain several duties. For example, developers still have to ensure that their designs are safe to construct and maintain. It is important to not overly depend on the DfSP.

The second misconception is that "the DfSP is expected to review all design risks and provide solutions". That is not true. The DfSP are delegated by the developer to oversee DfS review meetings and DfS register. The role is well-defined and focused on these two main duties. On the other hand, developers have to demonstrate that DfS is important and ensure the DfS review process is conducted effectively. In addition, designers are to identify, eliminate and reduce design risks during their design work. Thus, instead of the DfSP, the developer and designers are expected to constantly look for design risks and find ways to remove or reduce them.

Thirdly, the DfS review is conducted once every three to six months. That should not be the case. Designers should consider the design risks during their design work. They should not wait for the developer or DfSP to convene the DfS review meetings to start considering DfS issues. DfS reviews should be conducted frequently and progressively. It can be included as part of the regular design meetings so that it is a natural part of the project. Some organisations integrate DfS as part of regular design meetings and not a separate meeting or something that tags along. If it is treated as an unimportant element that is done only after all other items have been completed, everyone will be uninterested and not motivated to discuss the DfS issues. Thus, developers must send a clear message on the importance of DfS and that it is part of the design process. If the review meeting is once every three to six months, it is probably because the DfSP is paid based on meetings, and the developer wants to minimise cost. That is when the project team realises that the developer is not serious about DfS.

The fourth misconception is that there is no need to discuss a hazard during DfS review if there is a regulation or a code of practice that mandates controls that will reduce the risk. As discussed in the CROSS-UK (2021) safety alert, even if a control or design risk is codified, there is still a need to ensure compliance and be on the lookout for exceptions. Besides, it is important to consider soft hazards, like assumptions and errors. Thus, the DfS review team should make use of the DfS review process to check on possible Type B code-related design risks and Type C process and people design risks. The DfS review process should be part of the design process and designers should make full use of the DfS review process to prevent accidents and ill health across the structure's lifecycle.

The fifth misconception is that a DfS review is only concerned with the operations and the maintenance stages of the structure. In contrast, the contractor risk assessment will cover the construction and demolition stages. That is untrue. As discussed in Sec. 5.4, contractor RA and DfS review will overlap; it is unavoidable. The DfS review should aim to be comprehensive and not try to push all construction-related hazards to the contractor. The aim is to treat the risks as early as possible and through design changes and controls. This is highlighted in former Singapore Deputy Prime Minister Tharman Shanmugaratnam's 2014 speech, which highlighted that one of the key reasons for mandating DfS is the series of construction accidents in 2013. Thus, DfS is expected to help reduce construction fatalities, and developers and designers must consider ways to reduce the risk of construction accidents and ill health.

The last common misconception is that DfS review should only review unique hazards. Referring to the DfS flowchart (Fig. 5.2), this misconception means that the DfS review should only focus on detailed DfS review. That is not true. Sometimes, even though you have standard solutions from past projects or in codes, it does not mean that they are relevant or will be adequately implemented. It is important to make use of the DfS review process to check for exceptions or poor implementation of standard solutions. Designers must proactively consider what they can do to reduce downstream risk, even if the design risk is not "unique". In some situations, there could be better solutions that could replace the standard controls.

5.8 Challenges and success factors

According to Goh and Chua (2016), the top three perceived problems in practising DfS are client's cost concerns, contractors coming into the project too late, and inconsistency in

DfS reviews or checks. To address these problems, the client needs to provide the necessary motivation to ensure that DfS is effectively implemented. In addition, as discussed earlier, ECI will be useful in improving DfS. Concurrently, the use of integrated digital delivery tools will help designers better foresee possible hazards. Goh and Chua (2016) found that the success factors for a DfS review are:

- committed stakeholders who come to the DfS review meetings prepared and ready to contribute;
- an effective facilitator and coordinator for the DfS review process to ensure that all stakeholders are able to contribute to the review process;
- the design, WSH, and operational competency of stakeholders (not just the facilitator) so that hazards, elimination, and mitigation controls can be effectively identified;
- information is available and reviewed prior to the meeting;
- detailed documentation and tracking of follow-up actions by stakeholders.

In the end, the personnel overseeing site operations (e.g., the Qualified Person for Supervision in the Building Control Act and facilities managers) will have to ensure that the DfS measures are effectively implemented on-site and during the structure's operations.

From a systems thinking perspective, the successful implementation of DfS requires developers and designers to have a fundamental shift in their mental models. This is especially difficult because developers and designers may be uninterested in the WSH of workers downstream compared to financial and design issues that are more pressing and directly affect them. The main archetype that will be applicable to the successful implementation of DfS would be Eroding Goals and, as advised in Chap. 3, it is important for the regulators, who are beyond market influences, to set the vision clearly through regulations, standards and guidelines. This vision must be held consistently and constantly communicated with the industry until it becomes a shared vision. There must be an open dialogue to listen to the challenges and confusion of implementing DfS, which are expected, and resources must be provided to improve the industry's competency in addressing design risks. At the same time, enforcement must be fair, strict and swift. These measures will help to close the performance gap between the desired and actual quality of DfS reviews in the industry. The vision of WSH risks being eliminated and mitigated upstream should not waiver.

Review questions

1. Give some examples of design risks and explain why they are classified as design risks.
2. What are the roles of the different stakeholders in DfS?
3. Discuss the differences between a contractor's risk assessment and a DfS review.
4. Explain the GUIDE process and the differences between GUIDE 1, 2 and 3.
5. Give examples of how different DfS tools should be used in a design-then-build project.
6. Create a sample design risk assessment form.
7. Examine Australia's Safe Design process. Explain how it can be used with the GUIDE process.
8. Give examples of how design changes can improve: (i) WSH during construction, (ii) WSH during maintenance, (iii) WSH during use of facilities and (iv) WSH during demolition.
9. If you are the DfS Professional appointed by a developer, how would you ensure that a DfS review meeting is effective?
10. Give examples of Type A, B and C design risks.
11. How can the developer prevent the six misconceptions of DfS?
12. How can a developer ensure that DfS is successfully implemented in its projects?
13. Using the eroding goal archetype, describe how a DfS review process can deteriorate in terms of quality.

References

Behm, M. (2005). Linking construction fatalities to the design for construction safety concept. *Safety Science*, **43**(8), 589–611.

Bluff, L. (2003). Regulating safe design and planning of construction works — A review of strategies for regulating OHS in design and planning of buildings, structures and other construction projects. Sydney: National Research Centre for OHS Regulation.

Gambatese, J. A., Behm, M., and Rajendran, S. (2008). Design's role in construction accident causality and prevention: Perspectives from an expert panel. *Safety Science*, **46**(4), 675–691.

Goh, Y. M. and Chua, S. (2016). Knowledge, attitude and practices for design for safety: A study on civil & structural engineers. *Accident Analysis & Prevention*, **93**, 260–266.

Health and Safety Executive. (2015). The construction (design and management) regulations 2015. http://www.hse.gov.uk/construction/cdm/2015/index.htm (accessed: 22 July 2015).

Hossain, M. A., Abbott, E. L. S., Chua, D. K. H., Nguyen, T. Q., and Goh, Y. M. (2018). Design-for-safety Knowledge Library for BIM-integrated safety risk reviews. *Automation in Construction*, **94** (October), 290–302. https://doi.org/10.1016/j.autcon.2018.07.010

Institution of Engineers, Singapore. (2018). Technical articles — Health and safety engineering. https://www.ies.org.sg/Publication/Technical-Resources (accessed: 29 November 2023).

López-Arquillos, A., Rubio-Romero, J. C., and Martinez-Aires, M. D. (2015). Prevention through Design (PtD). The importance of the concept in engineering and architecture university courses. *Safety Science*, **73**, 8–14.

Occupational Safety and Health Council. (2016). Worked examples of design for safety. https://www.devb.gov.hk/filemanager/en/content_29/Design_for_Safety_Worked_Examples.pdf (accessed: 11 June 2019).

Real Estate Developers' Association of Singapore. (2019). DfS & WSH good practice guide. http://www.redas.com/goodpracticeguide.html (accessed: 11 March 2020).

Safe Work Australia. (2018). *Safe design of structures*. Canberra: Safe Work Australia.

Schulte, P. A., Rinehart, R., Okun, A., Geraci, C. L., and Heidel, D. S. (2008). National prevention through design (PtD) initiative. *Journal of Safety Research*, **39**(2), 115–121.

Szymberski, R. (1997). Construction project safety planning. *TAPPI Journal*, **80**(11), 69–74.

Toh, Y. Z., Goh, Y. M., and Guo, B. H. W. (2016). Knowledge, attitude and practice of design for safety: A study on multiple stakeholders in the construction industry. *Journal of Construction Engineering and Management*, **143**(5), 04016131.

UK Collaborative Reporting for Safer Structures. (2021). CROSS safety alert: The management of design related risks: Structural civil and fire engineers. December 2021. https://www.cross-safety.org/sites/default/files/202112/cross_safety_alert_the_management_of_design_related_risks.pdf

UK Design Best Practice. (n.d.). http://www.dbp.org.uk/ (accessed: 11 March 2020).

WorkCover New South Wales. (2001). *CHAIR — Safety in design tool*. Sydney: WorkCover New South Wales.

Workplace Safety and Health Council. (2015). Design for safety. https://www.wshc.gov.sg/ (accessed: 12 October 2015).

Workplace Safety and Health Council. (2016). Workplace safety and health guidelines — Design for safety. https://www.wshc.sg/files/wshc/upload/cms/file/WSH_Guidelines_Design_for_Safety(1).pdf (accessed: 6 July 2017).

Workplace Safety and Health Council. (2022). Workplace safety and health guidelines — Design for safety. https://www.wshc.sg/files/wshc/upload/cms/file/WSH_Guidelines_Design_for_Safety(1).pdf (29 November 2023).

Appendix: Sample Design Risk Assessment Form

DfS Risk Assessment Form												
Project Title:												
Company:												
Review:						Conducted by:						
Review Date: Next Review Date: Process/Location:												
S/N	Design Consideration	Hazards	Risk Assessment			Can these hazards be designed out?	Proposed Control Measures	Residual Risk Assessment			Further Review Required?	Action By
			Severity	Likelihood	Risk			Severity	Likelihood	Risk		

CHAPTER 6

Overview of workplace safety and health management systems

The Event Causation Technique

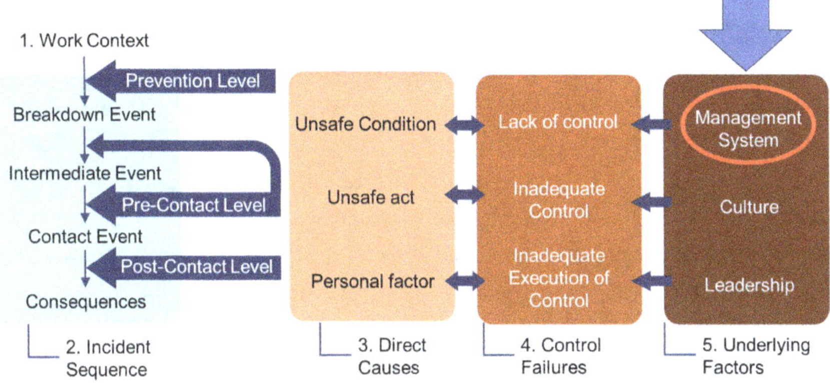

As identified in the Event Causation Technique (ECT) diagram above, the workplace safety and health (WSH) management system is an important underlying factor that can ensure suitable risk controls are implemented effectively to prevent incidents. WSH management systems are closely related to WSH culture and leadership, which will be covered in Chap. 7.

6.1 Introduction

This chapter introduces workplace safety and health (WSH) management systems. The Singapore Standard (SS) ISO 45001: 2018 (2023) Occupational Health and Safety Management

Systems — Requirements with Guidance for Use (Singapore Standards Council, 2023) will be used as an example of a WSH management system framework. SS ISO 45001:2018 (2023) is essentially ISO 45001:2018 as the Singapore Standards Council adopted the ISO standards for use in Singapore. It must be noted that there are other WSH management system frameworks, for example, SS 679:2021 Code of Practice for Workplace Safety and Health Management System for Construction Worksites (Singapore Standards Council, 2021), ANSI/ASSE Z10-2012 (R2017) (American National Standard (ANSI)/American Society of Safety Engineers (ASSE), 2017), and the International Labor Organization (ILO) guidelines on occupational safety and health management systems (International Labor Organization, 2001). There are also numerous WSH management system frameworks developed by different companies, such as ExxonMobil's Operations Integrity Management System (OIMS) (ExxonMobil, n.d.) and DNV's International Sustainability Rating System™ (ISRS) (DNV, n.d.). Many of the management system frameworks developed by companies integrate WSH with other areas such as environmental, quality, and security management.

6.2 Management system as an underlying factor

According to Definition 3.10 of ISO 45001:2018, a management system is a "set of interrelated or interacting elements of an organization (3.1) to establish policies (3.14) and objectives (3.16) and processes (3.25) to achieve those objectives". (The numbers in parentheses refer to corresponding definitions in other clauses in the standard). The term elements include "the organization's structure, roles and responsibilities, planning, operation, performance evaluation and improvement". Based on this definition of a management system, occupational health and safety (OHS)[1] management system is defined as a "management system or part of a management system used to achieve

---

[1] ISO 45001:2018 uses OH&S or OHS, which has the same meaning as WSH. The terms will be used interchangeably.

OHS policy". In contrast, OHS policy refers to the "intentions and direction of an organization (3.1), as formally expressed by its top management (3.12)".

As can be seen, according to ISO 45001:2018, a management system is essentially a set of fundamental management processes and structures. The overall purpose of a management system is to help an organisation set and achieve its policies and goals and continually improve itself. The basic Plan–Do–Check–Act (PDCA) cycle is fundamental to a management system (see Chap. 1). One key purpose of the different elements is to help managers execute and maintain the PDCA cycle.

Understanding the difference between a management system (an underlying factor in ECT) and a specific control measure (related to the "control failure" block in ECT) is important. A control can be a barricade, a training course, safety rules, or personal protective equipment (PPE). This will typically manage the direct causes (e.g., unsafe acts, unsafe conditions, and personal factors), and these are usually event-level remedies (see Chaps. 1 and 3) that contrast a management system intervention, which is more fundamental and general, e.g., consultation process, risk assessment process, and process to determine legal requirements. Control measures are usually identified through risk assessments (RAs) and are meant to reduce the risks posed by a hazard. In contrast, a management system influences the patterns and trends of unsafe acts and conditions in the workplace.

Let us take, for example, a worker working unsafely at height on wooden planks that are too flexible and unstable. An event-level answer to the problem would be to identify the planks as unsafe and then change them to ensure that the worker is safe. However, a more systemic approach would be to identify that the planks are unsafe, then determine if the same type of planks are widely used in the organisation and whether workers can identify the danger they pose to workers working on them. Suppose the planks are used on all the sites in the organisation, and workers cannot perceive the planks as hazards. In that case, from a systems point-of-view, there is a dangerous pattern indicating inadequacies in the management system, such as, RA process (ISO 45001:2018 Clause 6.1.2), procurement process (Clause 8.1.4), and evaluation of training effectiveness (Clause 7.2). Using the procurement process as an example, it may be discovered that the process did not include WSH criteria and feedback from site personnel before procurement of the planks and other work equipment and material. The procurement department may have been overly focused on price and did not have the competency to evaluate WSH criteria. The procurement process can be improved by requiring experienced site personnel to provide feedback on purchase specifications and require procurement staff to have basic WSH knowledge. An electronic feedback form can be developed, and relevant site personnel should be identified as subject matter experts for different types of purchases.

The above example shows that by improving the management system elements, in this case, the procurement procedure, the organisation as a whole becomes safer — not only will the planks used by workers on site be safer, but all newly purchased equipment and material should be safer due to input from experienced site personnel. Thus, ECT identifies management system as an important underlying factor that interacts with culture and leadership to influence the effectiveness of all WSH controls and direct causes that can lead to incidents.

6.3 Overview of ISO 45001:2018

The following will provide an overview of ISO 45001:2018 as an illustration of the key elements of a WSH management system. Readers are encouraged to read the actual standard for more details.

Figure 6.1 shows the management system framework of ISO 45001:2018. This framework is aligned with other ISO

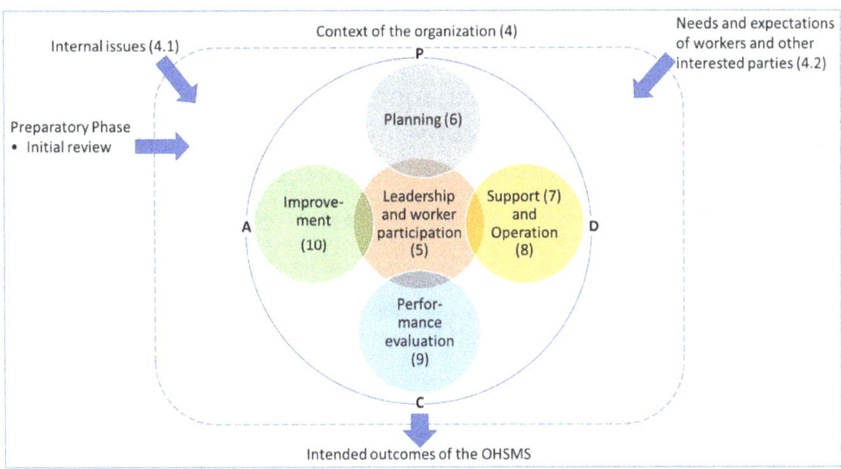

Fig. 6.1   ISO 45001:2018 Management system framework. (Adapted from (ISO, 2018)).

management system standards, such as ISO 9001 (quality management) and ISO 14001 (environmental management). Each of the elements in the framework will be discussed in the following sections. Note that the "initial review" element was inserted into the recommended framework by the author.

6.3.1 Initial review

With or without a formal management system, organisations have their way of managing themselves. These processes may or may not be documented, but they are happening. When an organisation is trying to set up a formal and documented management system, they start with an initial review, which compares existing management processes with the processes stipulated in the management system standard that they are trying to be certified to or model after, e.g., ISO 45001:2018 or SS 679:2021. As part of the initial review, existing risk management (RM) documents will have to be reviewed, and additional RAs may have to be conducted to understand the WSH risks that the organisation faces. A thorough check against legal requirements will also have to be conducted to understand the demands of the relevant legislation. Existing WSH training, programmes and initiatives should be documented. The resources used to conduct existing management system processes, programmes, etc., must be documented. Once the current management approach is documented, it can be compared against the WSH management system standard to identify gaps so that specific actions can be identified and implemented to close them.

6.3.2 Success factors

The ISO 45001:2018 Clause 0.3 highlighted a list of success factors for an effective WSH management system. Table 6.1 classified the 11 factors into five categories.

Table 6.1   Categorisation of success factors identified in ISO 45001:2018.

Category	Success factors identified in ISO 45001:2018
(I) Top management, culture, and resource allocation	a) Top management leadership, commitment, responsibilities and accountability
	b) Top management developing, leading and promoting a culture in the organisation that supports the intended outcomes of the OHS management system
	c) Allocation of the necessary resources to maintain it
(II) Policies, objectives and compliance	a) OHS policies, which are compatible with the overall strategic objectives and direction of the organisation
	b) OHS objectives that align with the OHS policy and take into account the organisation's hazards, OHS risks, and OHS opportunities.
	c) Compliance with its legal requirements and other requirements

Table 6.1   (*Continued*)

Category	Success factors identified in ISO 45001:2018
(III) Communication	a) Communication
	b) Consultation and participation of workers and, where they exist, workers' representatives
(IV) Effective and integrated processes	a) Effective process(es) for identifying hazards, controlling OHS risks, and taking advantage of OHS opportunities
	b) Integration of the OHS management system into the organisation's business processes
(V) Continual improvement	a) Continual performance evaluation and monitoring of the OHS management system to improve OHS performance

First, for a WSH management system to be successful, the top management needs to provide WSH leadership, build a strong safety culture, and provide sufficient resources for WSH management. Second, the top management must lay down clear WSH policies and objectives that guide the implementation of the WSH management system and ensure compliance with legal and other requirements. Third, the policies and objectives must be communicated with employees and stakeholders, and opportunities for consultation and participation on WSH matters must be provided. Fourth, the WSH management processes must be effective and integrated into the organisation's business processes, especially the RA process. Last, there must be continual improvement processes to improve WSH performance.

As can be observed, the key success factors are aligned with the ECT, which highlights the importance of leadership, culture and management systems. The 11 success factors are useful reminders to managers of what to focus on when managing a WSH management system.

6.3.3 Context

A WSH management system does not exist in a vacuum. The context in which it exists contains different issues that will affect how the WSH management system should be designed, operated and maintained. These issues can be classified into internal and external issues. BS 45002:2018 (BSI Standards Limited, 2018) provides further guidelines on possible internal and external issues. External issues include relationships with external providers such as contractors or suppliers, new technologies, cultural, social and political factors, and legislation. Internal issues include the size, nature and activities of the organisation, how the organisation is managed, its business objectives, resources, knowledge and competence, and planned or foreseeable changes and how these are managed.

When identifying relevant issues in the context, BS 45002: 2018 (BSI Standards Limited, 2018) recommended approaches such as "what-if" questions for organisations that run less complex and smaller operations. For organisations running more complex and larger operations, structured methods such as SWOT (Strengths, Weaknesses, Opportunities, and Threats) or PESTLE (Political, Economic, Social, Technological, Legal, Environmental) analysis can be used.

One key internal context that must be established is the needs and expectations of workers and other interested parties. Thus, Clause 4.2 of ISO 45001:2018 requires the organisation to identify workers ("workers" include managerial, supervisory and non-managerial staff) and other interested parties who are relevant to the WSH management system, their needs and expectations, and the relationships between these needs and expectations and relevant legal requirements.

Clause 4.3 specifies that the scope of the WSH management must be determined. The scope, i.e., boundaries and applicability, can be determined based on the types of activities, locations, products, and services the organisation has. For example, an organisation may exclude certain locations from the WSH management system initially because the location's operations are rather simple and pose minimal WSH risks.

6.3.4 Leadership and worker participation

6.3.4.1 Leadership and commitment (Clause 5.1)

Unlike its predecessor, OHSAS 18001:2009, ISO 45001:2018 focuses on leadership and worker participation. Clause 5.1 stipulates 13 sub-clauses on the top management's responsibilities. These include taking overall responsibility and accountability for WSH, ensuring the establishment of the WSH management system, integrating the WSH management system into the organisation's business processes, ensuring adequate resources, communicating the importance of effective WSH management and conformance, etc. The responsibilities are essentially elaborations of the 11 success factors highlighted earlier and top management's responsibilities over processes highlighted in other parts of SS ISO 45001:2018 (2023).

6.3.4.2 WSH policy (Clause 5.2)

WSH policy is one of the most fundamental elements of a WSH management system. It is essentially a statement that top management issues to demonstrate and communicate their commitment to WSH. However, despite its theoretical importance, the policy statement does not directly impact an organisation's WSH, and it is becoming too easy for organisations to employ consultants to help them draft suitable policy statements. The most important part of Clause 5.2 is perhaps the phrase "[to] establish, implement and maintain" — especially the word "implement". It is important for the policy, the key guidance for the rest of the management system, to be implemented. To be implemented means that the management system does not just exist on paper but is being used by people on the ground. "Maintained" means it is being utilised, checked and improved sustainably. This requires active effort on the part of the organisation. Many systems start well but deteriorate due to lack of maintenance. Many of the elements of ISO 45001:2018 (such as performance evaluation and improvement) are designed to ensure the active maintenance of the system.

6.3.4.3 Organisational roles, responsibilities and authorities (Clause 5.3)

The top management is required to assign responsibilities and authorities for relevant roles in the WSH management system. This can include roles such as a management representative overseeing the WSH management system, RA teams, WSH officers, WSH committee members and WSH internal auditors. The roles, responsibilities and authorities must be documented and communicated to all interested parties. Workers (including managers and supervisors) shall assume responsibility for those aspects of the WSH management system over which they have control. However, ISO 45001:2018 clearly states that top management is still accountable for the functioning of the WSH management system.

One of the most basic ways to communicate the roles, responsibilities and authorities is to have an organisation chart and accompanying descriptions of the relevant WSH roles, responsibilities and authorities for each position in the organisation chart. In addition, for each process or project, the roles and responsibilities can be described based on "RACI", which stands for Responsible, Accountable, Consulted and Informed. "Responsible" refers to managers or persons who implement the work or process. "Accountable" refers to the owner or champion for the work or process who signs off or approves the work or process. The person accountable will take overall responsibility for the work or process. "Consulted" refers to interested parties that must give inputs to the work or process to enable effective implementation. "Informed" refers to people who need to be kept in the picture but do not need to be formally consulted and are not directly involved in the work or proess.

6.3.4.4 Consultation and participation of workers (Clause 5.4)

ISO 45001:2018 requires a process (or processes) for the consultation and participation of workers (and their representatives) at all levels and functions. Participation is defined as involvement in decision-making, while consultation is defined as seeking views before deciding. The necessary resources, training and time must support the consultation and participation mechanisms. In addition, management must determine and remove or minimise obstacles or barriers to participation. Some of these barriers and obstacles include failure to respond to worker inputs or suggestions, language or literacy barriers, reprisals or threats of reprisals, and policies or practices that discourage or penalise worker participation.

ISO 45001:2018 emphasises the consultation of non-managerial workers on issues such as WSH needs and expectations, WSH policy, assignment of roles and responsibilities, WSH objectives and plans, controls for outsourcing, procurement and contractors, WSH monitoring, measurement and evaluation, audit, and continual improvement. Regarding the participation of non-managerial workers, the following are emphasised: mechanisms for consultation and participation, RA and controls, competency issues, communication processes, incident investigation, and corrective actions.

6.3.5 Planning (Clause 6)

The organisation must create plans to achieve the policy and desired outcomes of the WSH management system, i.e., preventing incidents and reducing WSH risks. The planning element consists of actions to address WSH risks, opportunities (Clause 6.1), objectives, and planning to undertake them (Clause 6.2). Clause 6.1 focuses on the RA process, similar to the RM process covered in Chap. 4. The management system will have to specify the procedure for the RM process. Organisations will have to take reference from relevant national standards or guidelines on RA and RM. However, there could be differences between the terminology used in ISO 45001:2018 and national standards or guidelines. For example, in the Singapore Risk Management Code of Practice (RMCP) (see Chap. 4), RA refers to hazard identification, risk evaluation, and risk control, but in ISO 45001:2018, RA is not clearly defined, and hazard identification and assessment of risks are separate. Putting the differences in terminology aside, both emphasise the importance of conducting a systematic RA (hazard identification, risk evaluation, and risk control).

It must be noted that ISO 45001:2018 (Clause 6.1.2.3) highlights the need to have a process (or processes) in place to assess WSH opportunities (in contrast to hazards and risks) to enhance WSH performance. The standard also requires considerations such as design of work areas, processes, installations, machinery/equipment, operating procedures, and work organisation, which aligns with the Design for Safety concept (see Chap. 5).

Clause 6.1.3 highlights the importance of having a procedure for identifying and accessing the legal and other WSH management requirements that apply to the organisation. This may seem straightforward, but due to the wide range of applicable legislations and standards, it is difficult to keep updated on the changes and evaluate the possible impact of these legislations and standards. Thus, many organisations subscribe to information services to keep themselves updated.

Clause 6.1.4 requires organisations to plan for actions to address risks, opportunities, legal, and other requirements and prepare for and respond to emergencies (Clause 8.2). The plan must be integrated with relevant processes and are implemented. They must also be evaluated for their effectiveness.

Clause 6.2 is concerned with objectives. Policies identify issues of great importance to the organisation; objectives are created based on the policy statement and WSH plans. For instance, if the organisation is concerned with the exposure to chlorinated solvents (used in degreasing and cleaning activities but are carcinogenic), then they can set an objective to eliminate the use of chlorinated solvents in all manufacturing lines by December 2026. The objective can also be supported by more specific targets, e.g., replacing 50% of chlorinated solvents in all manufacturing lines by June 2025 and replacing 75% of chlorinated solvents by September 2025.

Once a set of objectives and targets has been determined, the organisation can have a plan which identifies:

a)   what will be done;
b)   what resources will be required;
c)   who will be responsible;
d)   when it will be completed;
e)   how the results will be evaluated, including indicators for monitoring;
f)   how the actions to achieve OHS objectives will be integrated into the organisation's business processes.

A series of activities, campaigns and events that help the organisation achieve its policies, objectives and targets will be contained in the plan. The plan must have an identified timeline and allocation of responsibilities and accountabilities.

6.3.6 Support (Clause 7)

To support the functioning of the different elements of the WSH management system, ISO 45001:2018 identified five areas of support: resources, competence, awareness, communication, and documented information. Resources are self-explanatory and have been highlighted in earlier clauses. Competence is critical in WSH, and workers, including managers and supervisors, must be trained to perform their work safely. Competence goes beyond mandatory training required by the government. Mandatory training often only provides the basics; customised training can be more critical. Thus, a training needs analysis must be conducted based on the RAs, applicable legislations, WSH standards, competency standards, and other relevant documents. Besides training, experience is also critical in ensuring competence.

Besides competence, workers need to be aware of a range of WSH management system information, such as the policy and objectives, benefits of improved WSH performance, consequences of poor WSH, incidents and investigation outcomes, RA outcomes and actions, and their right to stop work when the WSH risk is not acceptable.

Communication refers to the communication processes needed for internal and external WSH communications. The processes must include the determination of the following:

(a) What needs to be communicated
(b) When to communicate
(c) With whom to communicate (internal, contractors, and interested parties), and
(d) How to communicate

The communication processes must consider possible barriers like literacy levels, language and cultural diversity, and disabilities.

Documentation is a fundamental process in a formal management system, but the level of documentation must be proportional to the operations' risk level, complexity and size. Legal requirements are also a major factor to consider when considering documentation. For instance, all RAs must be documented in Singapore and many countries. At the same time, the control of documented information needs to be robust to ensure the accessibility and reliability of the information. A common problem in organisations is the failure to ensure the proper versioning of documents, leading to outdated documents being used at the workplace. Processes for document control are critical to prevent miscommunication and the application of outdated procedures.

6.3.7 Operation (Clause 8)

Clause 8 covers the processes needed to prevent WSH incidents and relates to the plans and actions identified during planning. Operational planning and control cover elements such as determining necessary process criteria that define safe operations, safety work procedures, and work instructions. The controls are identified based on the RA, and those that are similar across different RAs can be grouped as general control measures, e.g., basic WSH rules, hazardous chemical procedures, programmes for the maintenance and repair of facilities, housekeeping procedures, traffic management plans, occupational health programmes, permit-to-work systems, and PPE programmes. High-risk activities that present specific hazards will require a dedicated set of controls and procedures for the activities based on the hierarchy of control discussed in the earlier chapter on RM.

An important process during operation is management of change (MOC) (Clause 8.1.3). MOC is implemented whenever there are changes in the workplace that influence WSH. These changes can be temporary (e.g., a truck breaking down at a construction site during a casting operation) or permanent (e.g., a new pressure vessel on the factory floor). The types of changes highlighted in ISO45001:2018 include:

- New products, services and processes, or changes to existing products, services and processes, including workplace locations and surroundings, work organisation, working conditions, equipment, and workforce.
- Changes to legal and other requirements.
- Changes in knowledge or information about hazards and risks.
- Development in knowledge and technology.

The organisation needs to define processes to evaluate these changes, assess the WSH risks and opportunities, and implement controls to manage the risks and opportunities. Organisations need to consider possible unintended consequences of changes. One example is when personal mobility devices (PMDs) were introduced to cut down walking time in a workplace, it unintentionally led to an increase in PMD collision accidents and fire incidents during the charging of the PMDs. Thus, relevant existing RAs have to be identified and reviewed.

Procurement (Clause 8.1.4) focuses on establishing WSH requirements for products and services to be purchased. The purchasing department will have to identify the standards that purchases have to comply with and write up relevant WSH specifications, such as required certifications, expectations for the WSH performance of the service provider, and required audits on the service provider or supplier. The procurement department will have to inform the supplier or service provider what is expected of them regarding their WSH management systems. The purchasing organisation may expect the supplier to be involved in the organisation's WSH activities, e.g., participate in safety campaigns with a similar level of training and RA. Pre-approvals of requirements, specifications and procedures for purchasing chemicals, machinery, and equipment will require the involvement of relevant experts and a clear process for checks before purchase. Suppliers' WSH performance and the goods they have supplied should also be recorded and evaluated. This information can then be used for future purchase evaluation. Goods, equipment and services should be inspected and verified regarding the WSH specifications and documented for future purchases.

Another important area is control over contractors. It is common for organisations to use contractors in their work. There are advantages to using contractors with the expertise that the organisation lacks, but organisations need to manage the contractors carefully. Contractors can be unfamiliar with the work environment, WSH requirements, and hazards within the client organisation, and they may have a weak WSH management system that do not meet the client's standards. Thus, it is important to establish clear criteria to select contractors demonstrating their capability to manage WSH. To do this, many clients rely on WSH certifications to standards like ISO 45001:2018 and other safety schemes or standards. In Singapore, experience shows that relying only on certifications and other schemes like bizSAFE[2] may not be the best gauge of good WSH performance. A mix of information like interviews, audit reports, client's experience with the contractors, and references from other clients is critical in the contractor selection process. Accident and incident records are useful sources of information, but many companies are simply lucky and serious accidents may be too rare to provide a good indication. Nevertheless, accident and incident records must still be reviewed when selecting contractors. Contractors, especially term contractors or those dealing with larger projects, must be evaluated and reviewed for their WSH performance and diligence in implementing their WSH management systems. Some basic checks would include ongoing inspections, review of method statements and RAs,

---

[2] A scheme under the Workplace Safety and Health Council, which is described as "a five-step programme that assists companies to build up their WSH capabilities so that they can achieve quantum improvements in safety and health standards at the workplace".

and review of safety and health committee meeting minutes. An important aspect is determining how senior and line management view WSH. If they see it as the WSH department's job, then it indicates that WSH is not taken seriously.

Suppliers and contractors are critical stakeholders in an organisation's management of WSH performance. Contractors frequently deal with highly hazardous processes. Due to the short-term nature of work, varying levels of risk perception, and other factors, contractors can cause WSH incidents. Clients, through their contracts and WSH department, will have to evaluate and monitor the WSH management activities of the contractors. This involves carefully selecting contractors based on their track records, WSH management system, and other evidence of their ability to manage WSH effectively. It is critical to store the organisational knowledge of different contractors and suppliers so that poor-performing contractors and suppliers will not be re-engaged. It is important to take a partnering approach, where the client helps the contractors and suppliers improve and uses incentives and recognitions to motivate them to progress in WSH management.

After the contractor has been selected, there must be frequent meetings to monitor how the contractor is progressing in its work and whether critical WSH procedures are implemented, e.g., RA, inspections, WSH meetings, incident investigation, and audits. The contractors' resources, such as plants, machinery, equipment, and chemicals, must be checked to comply with the company's requirements before they can be brought into the workplace. There can also be sharing incident learning and good practices among the contractors or subcontractors.

Outsourcing has similarities to the engagement of a contractor. The key difference is that the outsourced function or process can be performed in-house. Still, the organisation chose to focus on their core business and outsource non-core functions or processes. The key reasons for outsourcing are cost, resources, and lack of time. As in the case of engaging contractors, the outsourced companies must be carefully selected and monitored.

Organisations also need to identify and prepare for the emergencies that can occur in their workplaces. The processes to prepare for and respond to these emergencies must be established, implemented and maintained. These processes include establishing an emergency response plan (including first aid), training for relevant personnel, periodic tests and exercises, performance evaluations, and communication of the plan to workers and interested parties.

6.3.8 Performance evaluation (Clause 9)

The key elements under performance evaluation include monitoring, measurement, analysis and performance evaluation (Clause 9.1), and internal audit (Clause 9.2). Performance evaluation processes cover both quantitative and qualitative measures of WSH performance. These will be closely related to the policies, objectives and targets. At the same time, the measures will also be guided by the RA, especially the high-risk hazards and critical controls. The measures can also be split into proactive and reactive measures. Proactive measures monitor the implementation of controls and management activities, which are not dependent on incidents and accidents, i.e., reactive measures. Some proactive measures include inspections, audits, evaluation of training effectiveness, behaviour-based observations, and perception or safety culture / climate surveys. Reactive measures count the number and rate of incidents and accidents, e.g., accident frequency, accident severity, and fatality rates. Evaluation of compliance is directly related to the legal and other WSH requirements identified during planning. During implementation and operations, periodic audits and assessments should be conducted to determine if the organisation complies with all applicable legislation and standards.

An audit is important in evaluating compliance with the management system. It is different from an inspection, which is usually only focused on event-level issues. According to ISO 45001:2018, an audit is a "systematic, independent and documented process for obtaining 'audit evidence' and evaluating it objectively to determine the extent to which 'audit criteria' are fulfilled". There is a strong emphasis on evidence, and the evaluation needs to be thorough. An audit is usually based on the review of documents, interviews with employees, and physical inspections. The three processes produce evidence for triangulation so that the auditor can understand how well the management system is established, implemented, and maintained. The audits can be conducted internally or by an external auditor. In the Singapore construction industry, for example, the Building and Construction Authority (BCA) and the Ministry of Manpower (MOM) require safety and health audits on construction contractors. The BCA requires ISO 45001 certification for contractors with a grade of C2 and above. The contractor grading determines the government contract size for which the contractors can bid. Independent and external auditors are usually required for certification audits. These contractors need to be audited annually. In contrast, MOM requires six monthly independent and external audits for projects worth S$30 million and above. Projects with a contract sum of less than S$30 million must conduct internal reviews of the management system every six months. Experience has shown that audits can be compromised due to commercial pressure on the auditors. Auditors can be pressured by their paymaster, the contractors, to be more lenient. Thus, the auditors' professionalism and the contractors' WSH culture play an important role in the success of the audit regimes.

"Management review" refers to a high-level top management review of how effectively the management system is performing. It allows top management to make necessary changes to system elements like policies, resources and processes to improve the performance. The management can review a wide range of documents to get a sense of how established, implemented and well-maintained the WSH management system is. These documents include incident investigation reports, audit reports, and performance indicators derived from objectives and targets.

6.3.9 Improvement (Clause 10)

Incident investigation is one of the most fundamental improvement processes, and the organisation must respond strongly and seriously to any incident. This is covered in more detail in Chap. 3.

Besides incidents, nonconformity can be uncovered during operations or inspections, such that employees and contractors may be found to violate WSH rules or procedures. Corrective actions will be proposed based on investigations into the incidents or nonconformity; these should be recorded, and their effectiveness and implementation tracked. Corrective actions are focused on removing causes of nonconformity and preventing their recurrence. On the other hand, opportunities for continual improvement can also be identified during the investigation or operations. These opportunities are not linked to any specific nonconformity or incidents, but they do enhance WSH performance.

6.4 Management systems as balancing loops

The PDCA cycle used in ISO 45001:2018 is also known as the Deming cycle. The basic premise of the PDCA cycle is the balancing loop, as discussed in Chap. 2, which is contextualised in Fig. 6.2. During planning, different processes facilitate the determination of WSH policies, which are translated into measurable objectives and targets. At the same time, WSH controls are planned ("Plan") for and then implemented during operations ("Do"). Checks ("Check") are then conducted to measure the WSH performance and conformity to planned controls. The performance gap determined during checking becomes the basis for improvement actions ("Act") determined by management. However, there is a delay between improvement actions and measured performance because audits, inspections and checks are conducted periodically and are not comprehensive.

Figure 6.2 perfectly fits the systems archetype of "Balancing a process with delay" (see Table 3.4 of Chap. 3). Imagine a person trying to achieve the desired temperature of the shower water by adjusting the amount of hot water. However, there is a delay between his action and the temperature change. To make things worse, the duration of the delay is unknown. The longer the delay between his action and the temperature change, the more difficult it is to achieve the desired temperature. The person will either be having cold water or hot water. Similarly, to achieve a WSH objective, the management might put in too many resources or too much effort, possibly leading to impressive short-term WSH performance but at the sacrifice of other

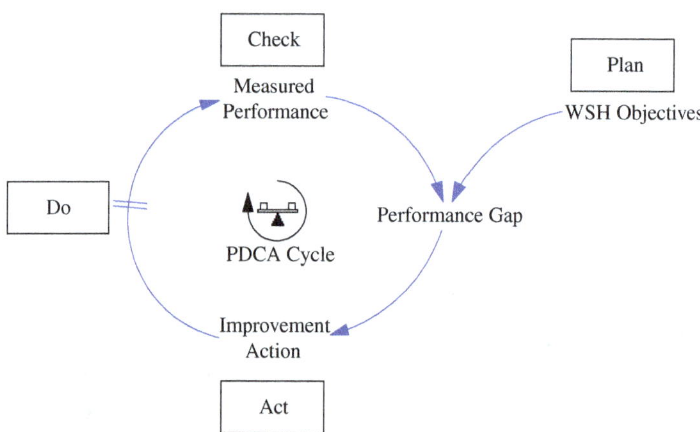

Fig. 6.2 PDCA cycle as a balancing loop with a delay.

organisational objectives. Or they might put in inadequate resources or effort, resulting in poor WSH performance.

Based on the advice in Table 3.4 of Chap. 3, the management needs to exercise patience, build an information system that gives quicker feedback, and actively seek the inputs of frontline staff. These are sound advice aligned with the guidance provided in ISO 45001:2018. Since WSH management systems are essentially control systems, other relevant systems thinking archetypes are "Eroding or drifting goals", "Fixes that fail", and "Shifting the burden (addiction)". Besides WSH management system standards, managers can refer to Table 3.4 for more insights to help them plan, establish, implement and maintain their WSH management systems.

6.5 Conclusions

The WSH management system manages the risk of WSH incidents. It is important for managers to think systemically and design processes and structures that ensure that suitable and effective controls are identified, implemented and maintained. This will then prevent the unsafe acts and conditions that cause incidents. ISO 45001:2018 is a comprehensive document that illustrates the key components expected of a robust and reliable WSH management system. From a systems thinking perspective, management systems are balancing loops constructed to help organisations achieve a desired level of WSH performance. The insights from different system archetypes are useful advice for managers trying to improve WSH performance.

Review questions

1. What is the difference between a risk control and an element or sub-system of a management system?

2. Based on the scenario described below, explain how the CEO's actions demonstrate the success factors identified in ISO 45001:2018.

    In a bustling construction company, BuildSafe Construction, CEO Samuel Goh is determined to implement ISO 45001: 2018. Samuel leads by example, frequently visiting sites to engage with workers and emphasize safety. After a near-miss incident involving falling debris, John spearheaded a new initiative to bolster the OHS management system.

    Sarah, the project manager, is tasked with aligning OHS policies with the company's strategic goals. She ensures resources are allocated to maintain the system, such as investing in new safety equipment and training programs. After a comprehensive hazard assessment, she sets specific OHS objectives to address site-specific risks.

    Clear communication channels are established, with regular safety meetings and an open-door policy for workers like Saravanan, a seasoned foreman, to voice concerns. This participative approach helps identify potential hazards and control risks effectively.

    The OHS management system is integrated into daily operations, with continual monitoring and evaluation driving improvements. For instance, after an evaluation, they implemented new safety protocols that significantly reduced slips and falls on site.

3. Give three examples of WSH Policies, Objectives and Targets for a large food and beverage outlet.
4. Explain how contractors and suppliers should be managed regarding WSH.
5. Discuss how performance evaluation using proactive and reactive measures are related to the proactive and reactive loops in Fig. 3.1 or Chap. 3.
6. What are the potential challenges in implementing a WSH audit system for a construction company, and how can they be mitigated?

7. Explain what management review is and why it is important.
8. Describe how the systems archetypes of "Eroding or drifting goals", "Fixes that fail", and "Shifting the burden (addiction)" can be used to guide the WSH management efforts for a transport operator.

References

American National Standard/American Society of Safety Engineers. (2017). ANSI/ASSE Z10-2012 (R2017) Occupational health and safety management systems.

BSI Standards Limited. (2018). BS 45002-0:2018 Occupational health and safety management systems — Part 0: General guidelines for the application of ISO 45001.

DNV. (n.d.). ISRS™: For the health of your business — Best practice in safety and sustainability management. https://www.dnv.com/oilgas/international-sustainability-rating-system-isrs/ (accessed: 28 May 2024).

ExxonMobil. (n.d.). OIMS: A disciplined management framework | Exxon-Mobil. http://corporate.exxonmobil.com/en/company/about-us/safety-and-health/operations-integrity-management-system (accessed: 3 April 2018).

International Labor Organization. (2001). Guidelines on occupational safety and health management systems. http://www.ilo.org/wcmsp5/groups/public/---dgreports/---dcomm/---publ/documents/publication/wcms_%20publ_9221116344_en.pdf (accessed: 2 September 2020).

ISO 45001. (2018). *Professional Safety*, **63**(2), 53–55. http://libproxy1.nus.edu.sg/login?url=https://search-proquest-com.libproxy1.nus.edu.sg/docview/1993301613?accountid=13876 (accessed: 2 September 2020).

Singapore Standards Council. (2021). SS 679:2021 Code of practice for workplace safety and health management system for construction worksites. Singapore: Enterprise Singapore.

Singapore Standards Council. (2023). ISO 45001:2018 (2023) Occupational health and safety management systems — Requirements with guidance for use. Singapore: Enterprise Singapore.

CHAPTER 7

Safety culture and leadership

The Event Causation Technique

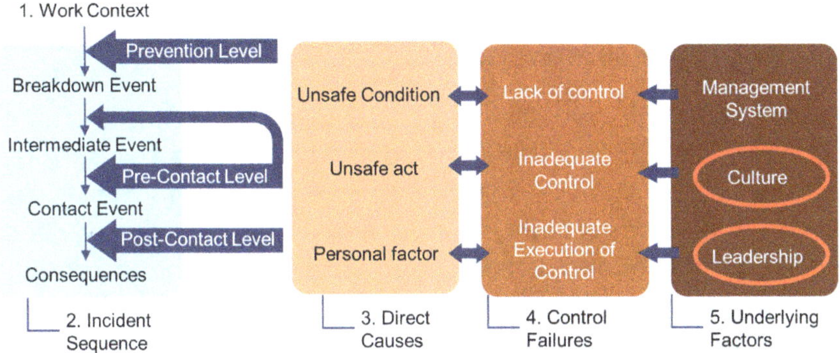

Safety culture and leadership are underlying factors within the Event Causation Technique (ECT) framework (see the ECT diagram above). They influence the effectiveness of the management system and the risk controls. Unlike management systems, which are usually documented, culture and leadership are harder to observe, evaluate and improve. Nevertheless, they are critical components of WSH management that have wide-ranging impacts.

7.1 Introduction

Culture affects all management processes in an organisation because it is the shared values and beliefs of people in the organisation. A Workplace Safety and Health (WSH) procedure or risk assessment (RA) may be very comprehensively written.

Still, it can be easily defeated if the workers on the ground and those in supervisory or management positions are not committed to them. In addition, leaders are the key people who have the greatest influence on safety culture. This chapter will first discuss the definition of safety culture by introducing several safety culture models. Subsequently, safety leadership will be introduced. Finally, the BP Texas City plant explosion case will be discussed to illustrate the importance of safety culture and leadership.

7.2 Defining safety culture

To define safety culture, we first need to understand organisational culture. One widely used definition of organisational culture is "a system of shared values *(what is important)* and beliefs *(how things work)* that interact with a company's people, organisational structures, and control systems to produce behavioural norms *(the way we do things around here)*" (Uttal, 1983) (text in italics added by author). Behavioural norms can also be seen as habits or habitual tendencies. Culture can be abstract because it comprises shared values and beliefs embedded in people's minds. However, culture can be observed through the pattern of behaviours across people at different times and locations.

Using the analogy of weather and climate, localised behaviours or specific events are like weather, which can vary in the short term. However, climatic patterns are relatively stable across longer periods and space. For instance, December is usually a rainy month in Singapore, but that does not mean that there is rain every day. The longer-term pattern of December being a wet month is similar to organisational culture, which is relatively stable across the years. However, there are occasional deviations from the general pattern, meaning that individuals may still exhibit values, beliefs and behaviours not aligned with the organisation's culture from time to time.

Following the same analogy, although the weather in Singapore is relatively uniform across the small island of 734.3 km² (about 50 km east to west, and 27 km north to south), the Meteorological Service Singapore has indicated that "rainfall is higher over the northern and western parts of Singapore and decreases towards the eastern part of the island". Likewise, organisational culture is not always uniform, even in small organisations. This is because each individual's values, beliefs and behaviours are always influenced by factors such as the person's family situation, experience, upbringing, personality, group influences, etc.

Applying the definition of organisational culture to safety, safety culture can be defined as a system of shared values and beliefs that interact with a company's people, organisational structures, and control systems to produce safety-related behavioural norms (see Fig. 7.1). The behaviour of each individual person in the organisation is affected by the individual's characteristics, the organisation's safety culture (shared values and beliefs), and the management system (structure and control system). Since the safety culture and management system similarly influence all individuals, there will be similarities in the behaviours of all individuals in the organisation. Still, some deviation is expected due to differences in individual characteristics.

This definition of safety culture is aligned with the Event Causation Technique (ECT), which highlights that risk controls, the WSH management system, safety culture, and leadership influence unsafe behaviour (a direct cause of incidents). Numerous researchers also highlight this concept, for example, Guldenmund (2007), who indicated that individuals' behaviours are influenced by the organisation's structure, culture and processes, which are "dynamically interrelated". He also indicated that an "organisation's culture cannot be isolated from its structure or processes".

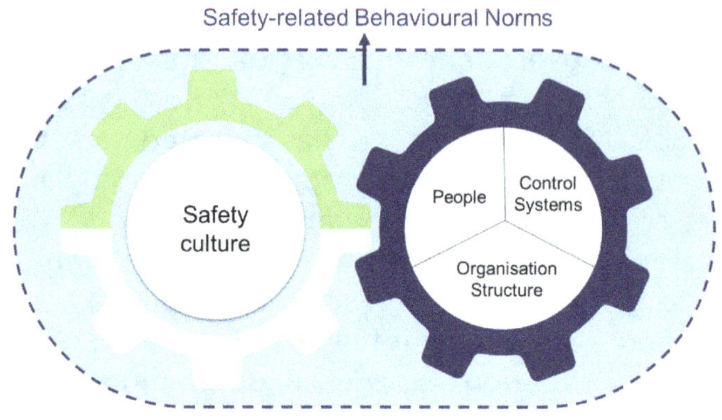

Fig. 7.1 Safety culture.

To understand safety culture further, it is important to define beliefs and values. Beliefs refer to basic assumptions that people hold in their minds. Some examples are, "safety is costly", "we have no time for safety and health issues", "accidents are a matter of luck", and "it all boils down to workers' behaviours; there is not much that the management can do". Beliefs drive behaviours. Hence, if one has beliefs contradicting sound WSH management principles, their decision-making and intentions can be unsafe. Values refer to the relative importance of different concepts, e.g., WSH versus productivity and integrity versus ambition. Work-related values would be influenced by the beliefs and assumptions we hold about the organisations we work in or with and the people in the organisations. Shared values and beliefs are a group's common values and beliefs. The common values and beliefs arise or are inculcated through shared experiences, stories and knowledge. The management system or norms in the group will help to create and maintain the culture in a workplace.

As culture is intangible, many managers do not focus on culture when managing the organisation, or they might not have the competency to influence culture. This can lead to management actions that fail to deliver or even backfire. It is important to design the management system to align with the desired culture while considering the existing culture.

7.3 Different layers of culture

Based on previous research in organisational culture, Guldenmund (2000) presented a multi-layer model of safety culture (see Fig. 7.2).

The core of safety culture is assumed to comprise shared basic assumptions, which are the shared beliefs discussed earlier. These assumptions include the nature of reality and truth, time, space, human nature, human activity, and relationships. They are mainly implicit but obvious or even factual to members of the organisation. Yet, they remain implicit, invisible and pre-conscious (i.e., assumed to be true in an unconscious fashion). These basic assumptions, or shared beliefs, have to be deduced from culture's middle and outer layers.

The middle layer is made up of espoused values or attitudes regarding hardware, software, people, and risks. Values or

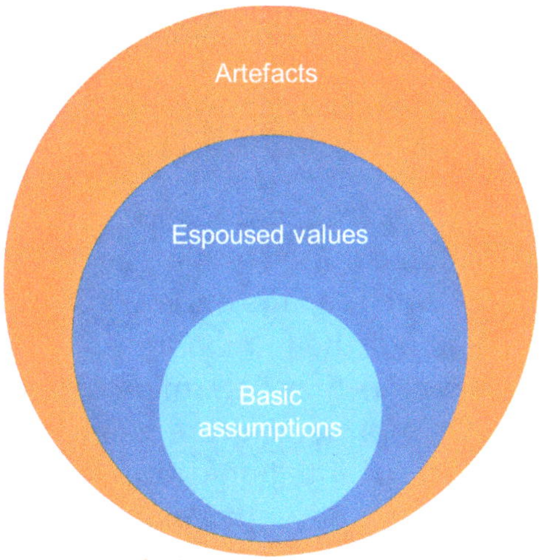

Fig. 7.2 Levels of culture.

attitudes arise from basic assumptions or beliefs. This layer is more explicit and conscious, and it includes people's attitudes towards objects or artefacts such as organisational policies, training manuals, procedures, formal statements, bulletins, accident and incident reports, job descriptions, and minutes of meetings. Finally, the outer layer refers to the objects or artefacts that are visible and representative of safety culture, such as those mentioned earlier. Other examples of artefacts include management policies, meetings, inspection reports, dress codes, personal protective equipment (PPE), posters, and bulletins. However, artefacts are poor representations of the underlying culture and are easily misinterpreted. For example, a company with many safety posters is not necessarily serious about the safety and health of its employees.

Guldenmund's multi-layer safety culture model highlights the difficulties in evaluating the safety culture of a workplace. Shared beliefs and values cannot be measured directly because many assumptions and beliefs are so deeply ingrained that individuals may not be aware of them. Managers will have to rely on measuring espoused values, attitudes and artefacts to judge the status of safety culture in their workplace. This will be discussed in more detail in this chapter and Chap. 8.

7.4 Reason's safety culture model

Reason (1997, chapter 9) argued that a positive safety culture is essentially an informed culture. To him, an informed culture is "one in which those who manage and operate the system have current knowledge about the human, technical, organisational and environmental factors that determine the safety of the system as a whole". An informed culture means the organisation is in a sustained state of intelligent and respectful wariness. The organisation gathers the right kind of data using a safety information system. They do not just rely on lagging indicators;

they also use leading indicators to help them manage safety and health risks proactively.

Reason (1997, p. 37) also highlighted, "… chronic unease is the price of safety. Studies of high-reliability organisations — systems having fewer than their 'fair share' of accidents — indicate that the people who operate and manage them tend to assume that each day will be a bad day and act accordingly". High-reliability organisations operate in highly complex environments and are expected to offer failure-free services. These organisations need to have robust WSH management systems supported by a positive safety culture and committed WSH leadership.

Reason (1997) felt that a safety culture can be socially engineered by implementing the different components of an informed culture. To him, an informed culture is based on four sub-cultures: reporting culture, just culture, flexible culture, and learning culture. All four sub-cultures interact and come together to produce an organisation that has an informed culture.

7.4.1 Reporting culture

Reporting culture refers to the involvement and willingness of people to report hazards, errors and incidents. Possible disincentives to reporting culture include concerns about the impact of reporting, and perceived additional work or trouble. To encourage reporting, there could be indemnity against disciplinary proceedings (as far as reasonably practicable) and ensuring confidentiality or de-identification of reporting staff.

Ideally, there should be an independent and credible department or agency for disciplinary actions, which is separate from the department receiving reports of hazards and making WSH improvements. Other incentives include prompt feedback to the community about the actions and assessment of reported hazards or incidents and ensuring the ease of making the reports. The aviation industry has implemented successful reporting programmes by implementing many of these incentives for reporting. The critical success factor for a reporting system is trust. Employees must understand the importance of reporting and feel safe reporting their errors.

7.4.2 Just culture

Just culture is essentially having a clearly established set of WSH rules, responsibilities and accountabilities, such that everyone in the organisation knows what is expected of them and accepts the consequences that come with violating the WSH rules. These rules are fair because they also consider the human factors involved in accidents and violations. Just culture is also about the way that the organisation handles blame and punishment. A "no-blame" culture is neither feasible nor desirable, as it may breed irresponsibility among some individuals, which can spread across the organisation. Conversely, a punish-no-matter-what approach is also not feasible.

Any punishment needs to consider the circumstances of the unsafe act. Even if the unsafe act was intentional, the worker committing the unsafe act need not be guilty. Managers must consider whether workers were encouraged to violate the rules due to inadequate supervision or ineffective training. It could also be a norm for everyone to break the rules because of management's lack of interest in WSH. Decisions on the punishment also need to take into consideration if the same unsafe act would be committed if another reasonable man was placed in the same situation. In addition, the worker's WSH track record needs to be factored in. These questions need to be considered prior to any punishment.

It is important to note that just culture promotes trust and helps engender a reporting culture. There must be a good balance between punishment and rewards because rewards are known to produce positive effects in developing safe behaviour. With reference to Table 7.1, immediate recognition or other rewards are usually better at producing greater motivation to be safe. Some examples include giving workers small gifts or badges during site inspections as recognition of their positive safety behaviour. This is more effective in motivating safe behaviour than scolding and punishment (doubtful effects).

Table 7.1.   Effect of reward/punishment on behavioural change. (Adapted from (Reason, 1997)).

	Immediate	**Delayed**
Reward	Positive effects	Doubtful effects
Punishment	Doubtful effects	Negative effects

7.4.3 Flexible culture

Flexible culture refers to the ability of an organisation to re-arrange its structure based on the needs of the work context.

A flexible culture allows people with the right expertise to provide input or take the lead when their expertise is needed. This is especially prominent in crises where the manager or supervisors may not have the expertise to manage the crisis. This may call for the fire safety manager or WSH manager to take over management of the site during the crisis. This can also happen during work, where those with different expertise are called upon to identify or control hazards. The ability to tap into expertise to suit the situation is a mark of a flexible culture.

One of the key elements of a flexible culture is the knowledge of where different expertise lies in the organisation. The safety information system needs to contain a comprehensive database of the strengths and competencies of different personnel within the organisation.

However, members of the organisation must also be willing to offer their expertise readily. Proactive involvement can be encouraged through incentives, but a sense of teamwork is more fundamental to a flexible culture.

7.4.4 Learning culture

Learning culture is the ability to learn from past incidents and available information to prevent future incidents. This involves a detailed analysis of information and having the ability to learn as a group. The ability to look beyond events and spot trends and patterns to prevent undesirable events is a key hallmark of a learning organisation. Learning culture requires an open mind and the willingness to listen to others.

However, an organisation need not have accidents to identify ways to improve their WSH management. The company can actively search for opportunities to learn by, for example, evaluating incidents from companies with similar operations, collecting feedback from workers periodically, reviewing audit reports, and analysing complaints from customers or other stakeholders. Another useful approach is to invite sister units or peer organisations to conduct benchmarking or peer review exercises to help the organisation review its performance and adopt best practices. Learning from successful departments or individuals within organisations and understanding their key success factors is also important. Improvement teams can be set up to innovate and focus on specific WSH issues. These improvement teams can be recognised and incentivised through competitions and awards.

7.5 Health and Safety Executive safety culture questionnaire

Safety culture comprises shared values and beliefs, which can be hard to measure and manage. Many researchers and organisations have developed questionnaires to help organisations assess safety culture. Many researchers use the term safety climate instead of safety culture because "safety climate" refers to the middle layer of safety culture (see Fig. 7.2), i.e., attitudes. In contrast, safety culture refers to the core, which is "preconscious" and hard to measure. Many questionnaires use a quantitative approach, but some use a qualitative approach, combining responses from site personnel and the inspector or auditor to determine safety culture improvement actions. One of these questionnaires is the Health and Safety Executive (HSE) safety culture questionnaire (Health and Safety Executive, n.d.), which is made up of seven key factors:

1. Management commitment
2. Communication
3. Employee involvement
4. Training/information
5. Motivation
6. Compliance with procedures
7. Learning organisation

In contrast to the four sub-cultures of Reason's model, the HSE safety culture questionnaire highlights the importance of management commitment and the need for managers to be seen as leading by example regarding safety. Another important aspect of management commitment is how managers allocate resources and how they make decisions when WSH and productivity or profit contradict each other. A good example of strong management commitment to safety is how Paul O'Neill, CEO of Alcoa (1987–2000), turned around the safety records of the company and how he, while maintaining these impressive safety records, made Alcoa one of the most profitable companies in the world (Duhigg, 2014; Roth, 2012).

Several of the HSE safety culture questionnaire components are aligned with Reason's model. A learning organisation is synonymous with learning culture in Reason's model. Communication, employee involvement, information, and motivation are related to reporting culture and just culture in Reason's model. However, the HSE safety culture questionnaire highlights the importance of training and compliance, which are not clearly highlighted in Reason's model. It is noted that some of the components of the HSE safety culture questionnaire, e.g., management commitment, communication, employee involvement, training and information, and compliance with procedures, are closely related to the WSH management system elements described in the previous chapter. This is not surprising because safety culture affects the way the WSH management system is established, implemented and maintained. A well-implemented WSH management system is reflective of a positive safety culture. The key difference is that the HSE safety culture questionnaire measures employees' perception of these WSH management system elements as a proxy of the WSH beliefs and values of the organisation.

7.6 Zohar's safety climate survey

Zohar (Hofmann *et al.*, 2017; Zohar, 1980; Zohar, 2010; Zohar and Luria, 2005) is known for his research on safety climate. Figure 7.3 shows an example of the questionnaire items in a typical safety climate questionnaire. The respondent will rate each statement on a Likert scale of 1 to 5 or 1 to 7. The aggregated score will then give the management a sense of the organisation's safety climate. In addition, it is possible to target specific management or supervisor behaviours to improve the safety climate. Zohar (2010) opined that safety climate is "a robust leading indicator or predictor of safety outcomes across industries and countries". The periodic measurement of an organisation's safety climate acts as a leading indicator, which must be read in conjunction with other leading and lagging indicators.

7.7 Design for Safety climate measurement tool

Research on construction safety climate has led to the creation of various tools to gauge safety climate in the construction

Organisation-Level Safety Climate
Top management in this plant–company...

1. Reacts quickly to solve the problem when told about safety hazards.
2. Insists on thorough and regular safety audits and inspections.
3. Tries to continually improve safety levels in each department.
4. Provides all the equipment needed to do the job safely.
5. Is strict about working safely when work falls behind schedule.
6. Quickly corrects any safety hazard (even if it is costly).
7. Provides detailed safety reports to workers (e.g., injuries, near accidents).
8. Considers a person's safety behaviour when moving/promoting people.
9. Requires each manager to help improve safety in his/her department.
10. Invests a lot of time and money in safety training for workers.
11. Uses any available information to improve existing safety rules.
12. Listens carefully to workers' ideas about improving safety.
13. Considers safety when setting production speed and schedules.
14. Provides workers with a lot of information on safety issues.
15. Regularly holds safety-awareness events (e.g., presentations, ceremonies).
16. Gives safety personnel the power they need to do their job.

Note. Items cover three content themes: Active Practices (Monitoring, Enforcing), Proactive Practices (Promoting Learning, Development), and Declarative Practices (Declaring, Informing).

Fig. 7.3   (*Continued*)

industry (Chen *et al.*, 2021). However, these tools are tailored to assess the frontline workers' perception of aspects like the contractor's safety commitment, safety communication between supervisors and workers, and use of personal protective equipment. They are not applicable to the developers and designers who do not work on the construction site. A new safety climate measurement tool must be developed to evaluate the safety climate of upstream stakeholders such as developers

Group-Level Safety Climate

My direct supervisor…

1. Makes sure we receive all the equipment needed to do the job safely.
2. Frequently checks to see if we are all obeying the safety rules.
3. Discusses how to improve safety with us.
4. Uses explanations (not just compliance) to get us to act safely.
5. Emphasises safety procedures when we are working under pressure.
6. Frequently tells us about the hazards in our work.
7. Refuses to ignore safety rules when work falls behind schedule.
8. Is strict about working safely when we are tired or stressed.
9. Reminds workers who need reminders to work safely.
10. Makes sure we follow *all* the safety rules (not just the most important ones).
11. Insists that we obey safety rules when fixing equipment or machines.
12. Says a "good word" to workers who pay special attention to safety.
13. Is strict about safety at the end of the shift, when we want to go home.
14. Spends time helping us learn to see problems *before* they arise.
15. Frequently talks about safety issues throughout the work week.
16. Insists we wear our protective equipment even if it is uncomfortable.

Note. Items cover three content themes: Active Practices (Monitoring, Controlling), Proactive Practices (Instructing, Guiding), and Declarative Practices (Declaring, Informing).

Fig. 7.3   Questionnaire items developed by Zohar and Luria (2005).

and designers. Consequently, Lim and Goh (2024) developed a Design for Safety (DfS) climate measurement tool (Table 7.2). The DfS climate construct consists of five dimensions and 16 items. The five dimensions are:

1. **Participation of leader**: This dimension represented whether the project's leadership took an active role during the DfS

Table 7.2 Items in the Design for Safety climate scale (Lim and Goh, 2024).

Dimension	Item
Participation of leader	The developer's representative(s) participates in the DfS review meeting
	The developer's representative(s) contributes ideas during the DfS review process
	The developer's representative(s) listens carefully to ideas raised during the DfS review process
	The developer's representative(s) prioritises the DfS review process over other commitments
	The developer's representative(s) reacts quickly to demands from the DfS review process.
Project resources	The developer's representative(s) gives sufficient time to the project to enable an effective DfS review
	The developer's representative(s) provides the project with sufficient resources to work on the DfS review process
Expectations for stakeholder representation	The developer's representative(s) expects the project team to attend the DfS review meetings
	The developer's representative(s) expects representations of different stakeholder perspectives during the DfS review meeting
Member participation	In my impression, members of the project team actively bring up potential hazards during the DfS review
	In my impression, members of the project team actively bring up potential mitigation measures during the DfS review meetings
	In my impression, members of the project team participate in the DfS review meeting
Member cooperation	In my impression, members of the project team are cooperative during the DfS review process

(*Continued*)

Table 7.2 (*Continued*)

Dimension	Item
	In my impression, members of the project team take responsibility to make the necessary design changes.
	In my impression, members of the project team respect the DfSP or the meeting facilitator
	In my impression, members of the project team cooperate with the DfSP's or the meeting facilitator's requests

review process and contributed to the meetings, helping to evaluate the DfS aspects of the project.

2. **Project resources**: This dimension is about perceptions of whether the project had adequate resources to execute the changes or review the design effectively.

3. **Expectations for stakeholder representation**: This dimension tapped on whether the leadership of the project team set an expectation for all stakeholders to be present at the DfS review. Stakeholders are those involved in the completed project's design, construction, or final use.

4. **Member participation**: This dimension represents the perception of whether team members have fully participated in the tasks during the review process and reviewed the design risks.

5. **Member cooperation**: This dimension represents the perception of the interactions between project team members and whether team members cooperate with other personnel in the task of DfS review.

Similar to other safety climate questionnaires, each of these items is measured using a 5-point Likert scale, and items under each dimension are aggregated to provide an overall score for each of the items. The score can then allow organisations to understand their DfS climate and areas for improvement.

7.8 Culture SAFE and others

Singapore's Workplace Safety and Health Council (2015) has a comprehensive safety culture assessment and improvement programme called CultureSAFE. CultureSAFE provides the safety culture diagnostic tool, which is a questionnaire survey and a range of recommended programmes to improve safety culture. Another well-known safety culture improvement programme is the Hearts and Minds programme (Energy Institute, n.d.), which Shell developed. The range of safety culture improvement programmes is very wide, and each has a different take on what safety culture entails. Nevertheless, the different models agree that safety culture is about human and organisational aspects of WSH management. Safety culture influences how the WSH management system and risk controls are established, implemented and maintained.

7.9 Safety leadership

Managers and leaders influence safety climate and safety culture through their decisions, actions and behaviours. This section will discuss the WSH roles of senior and project managers. Selected leadership models will also be discussed in the context of WSH.

One of the key leadership approaches adopted by many leaders to influence subordinates' safety behaviour is the use of reward and punishment systems. This is known as transactional leadership (Northouse, 2016). In contrast, transformational leadership "evokes changes in subordinates' value systems to align them with organizational goals" (Clarke, 2013). Numerous studies have shown that it has positive effects on the safety behaviour and safety participation of subordinates. This section will highlight how leaders can adopt the transformational leadership style so as to engender a positive safety culture in their workplace. Another popular leadership model relevant to our discussion is Situational Leadership® II (SLII®) (Blanchard, 1985; Blanchard *et al.*, 2013; Northouse, 2016), which will be discussed in this section.

7.9.1 Transformational leadership

Many leadership theories have been developed over the years, but since Downton (1973) coined the term "transformational leadership", it has become one of the most widely researched leadership theories (Northouse, 2016). So what is transformational leadership? It is widely accepted that transformational leadership encompasses the following distinct but correlated dimensions (aka the 4 'I's):

1. Idealised influence, i.e., leader instills confidence and behaves in admirable ways that make followers identify with them.
2. Inspirational motivation, i.e., leader inspires others towards goals, provides meaning, optimism and enthusiasm, and articulates a vision that is appealing and inspiring to others.
3. Intellectual stimulation, i.e., the leader challenges assumptions, takes risks, and encourages subordinates to be creative.
4. Individualised consideration, i.e., the leader shows interest in subordinates' personal and professional development and listens to followers' needs and concerns.

At the same time, much research has shown that WSH leadership is one of the key criteria for a positive safety climate or culture (Zohar, 2010). In essence, "leaders create climate" (Lewin *et al.*, 1939). To facilitate ease of discussion, a climate is deemed to be a proxy to culture, i.e., they are treated as "pretty much the same thing". In the case of transformational leadership, it is quite intuitive that transformative leadership will positively impact safety culture. A closer look at each of the 'I's in transformational leadership will shed more light on the relationship between transformational leadership and safety culture (adapted from (Clarke, 2013)):

1. Idealised Influence: Managers adopting transformational leadership are role models by doing what is morally right as compared to focusing on production goals only. Idealised influence encourages a focus on WSH and sustainable ways of working, in contrast to prioritising short-term benefits due to work pressure. Leaders high in idealised influence will demonstrate their commitment to WSH as a core value. Personal commitment will improve followers' trust in management and loyalty, which can lead to improvement in overall performance.

2. Inspirational motivation: Transformational leaders motivate followers to be part of the organisation's shared vision and go beyond their needs. In terms of WSH, transformational leaders inspire their followers to achieve safety levels previously believed to be impossible, e.g., Vision Zero or Zero Harm. Transformational leaders frequently use symbols and stories to articulate their vision and inspire followers.

3. Intellectual stimulation: Intellectual stimulation arises when leaders encourage followers to address WSH issues and enhance information sharing by challenging long-held assumptions and promoting innovation in WSH. This engenders a learning culture where followers become more aware of and better understand WSH problems, encouraging innovative rather than reckless solutions.

4. Individualised consideration: Transformational leaders have a strong and sincere interest in their followers' well-being, which naturally includes WSH. Transformational leaders are not satisfied with compliance with WSH legislation because

they want to ensure that their followers do not experience injuries and ill health.

Empirical studies support the positive impact of transformational leadership on safety culture and safety performance. For example, Barling *et al.* (2002) studied 174 restaurant workers and 164 young workers from diverse jobs using selected questions from the Multifactor Leadership Questionnaire (MLQ) (Bass and Avolio, 1990), a shortened version of the safety climate questionnaire by Zohar (1980). They also measured the self-reported safety consciousness and safety incidents. The study showed that, among other factors, safety-specific transformational leadership plays a significant role in influencing "safety climate, safety consciousness, and safety-related events". On the other hand, passive leadership is shown to be related to poorer safety climates (Kelloway *et al.*, 2006). Thus, past studies have shown that leadership influences climate, and in turn, climate has been shown to influence the occurrence of workplace injuries (Huang *et al.*, 2012). Transformational leadership is a potentially powerful tool to influence safety culture and WSH.

Based on the 4 'I's, leaders can adopt the following practices to be more aligned with transformational leadership (Kouzes and Posner, 2002):

1.   Model the way: Be clear about one's values and philosophy, and be a role model.
2.   Inspire a shared vision: Develop a compelling vision that helps followers visualise positive outcomes and see how their dreams can materialise.
3.   Challenge the process: Transformational leaders are pioneers because they are willing to experiment, try new things, and learn from mistakes to achieve better results.
4.   Enable others to act: Outstanding leaders effectively work with people by building trust, promoting collaboration,

and listening respectfully. They empower others to make decisions and feel good about their work by showing the connections with greater purposes.

5. Encourage the heart: Be attentive to followers' need for recognition and reward. Show appreciation and encouragement to others.

Transformational leadership can improve organisations' performance, including safety culture and WSH performance. Positive changes in followers' behaviour will take time, but the persistent implementation of transformational leadership will reap results in WSH and organisational performance in general. This is not to say that it is the only leadership style that applies to all situations. Successful leaders will have the wisdom to observe and apply suitable leadership actions in different situations to improve WSH. In the end, leaders must take it upon themselves to use transformational leadership or other approaches to convince their followers to adopt safer work methods.

7.9.2 Situational Leadership®

The basic premise of the SLII® model (Blanchard, 1985; Blanchard *et al.*, 2013; Northouse, 2016) is that leaders need to adapt their leadership style and behaviour based on the development level of their followers. The leadership dimension is simplified into a dichotomy of directive and supportive. The followers' development levels are categorised by their competency and commitment to assigned goals. Followers' competency and commitment vary across time and situations, so the leader must match the leadership approach to suit the followers' collective level of competency and commitment.

A directive leadership behaviour involves one-way communication of the goals, methods to conduct the task, ways to evaluate the performance, etc., from the leader to the follower. On the other hand, supportive leadership behaviour uses two-

Table 7.3 Leadership approach based on the SLII® model (Blanchard, 1985; Blanchard *et al.*, 2013; Northouse, 2016).

Leadership approach	Leadership style		Followers' development level	
	Directive	**Supportive**	**Competence**	**Commitment**
Directing	High	Low	Low	High
Coaching	High	High	Low to some	Low
Supporting	Low	High	Moderate to high	Variable
Delegating	Low	Low	High	High

way communication to show social and emotional support to followers.

Based on Table 7.3, leaders should use the suitable safety leadership approach when working with different types of followers. Suppose the followers do not have suitable competency in WSH management but are highly committed to improving WSH. In that case, the leader should focus on communicating WSH goals, giving detailed instructions on the methods to be used by the followers and then supervising them carefully (Northouse, 2016). Coaching is suitable when the follower has little or no competence and low commitment to WSH. Besides being directed, the leader needs to support the followers through consultation, solving problems together, and encouraging them. Supporting is used when the followers have moderate to high competence in WSH management but have variable commitment. In this situation, the leader tries to motivate the followers to contribute their best skills. The leader will have to listen, give feedback, and praise the followers appropriately. The followers will be directly in charge of the day-to-day implementation of WSH management tasks, but the leader will still be involved in goal-setting and problem-solving. Last, when the followers are highly competent and committed, the leaders aim to get the group to agree on the goals and directions and let followers be responsible for achieving the goals based on what they think is suitable.

The SLII® provides useful WSH leadership guidance to front-line leaders and top management. A supervisor can assess each worker based on competence and commitment to decide how to direct or support the worker. Similarly, from the perspective of an organisation, the top management will have to assess the level of WSH competency and commitment of its employees in general, which is an indication of the safety culture of the organisation, and then decide what leadership approach is needed to improve safety culture as a whole or use different approaches for different groups within the organisation.

7.9.3 Chief executives' and board of directors' workplace safety and health duties

WSH legislation should motivate leaders and managers to be proactive and interested in ensuring WSH in their organisations. In Singapore, Section 48(1) of the WSH Act states that companies' chief executives and board of directors (company directors) can be held liable if their company has committed an offence under the WSH Act. The company directors must prove that they have exercised their due diligence. This approach is similar to the UK's Corporate Manslaughter and Corporate Homicide Act 2007, in which companies and organisations can be found guilty of corporate manslaughter as a result of serious management failures resulting in a gross breach of a duty of care.

Another similar requirement is found in Section 27 of the Australian model Work Health and Safety (WHS) Act (Safe Work Australia, n.d.), which requires the officer of a person conducting a business or undertaking (PCBU) to exercise due diligence to ensure that the PCBU complies with the Act. The Australian model WHS Act Section 27(5) states that due diligence involves taking reasonable steps to:

1. Acquire and keep up-to-date knowledge of WHS matters
2. Gain an understanding of the nature of the operations of the business or undertaking of the PCBU and generally of the hazards and risks associated with those operations
3. Ensure that the PCBU has available for use and appropriate resources and processes to eliminate or minimise risks to health and safety from work carried out as part of the conduct of the business or undertaking
4. Ensure that the PCBU has appropriate processes for receiving and considering information regarding incidents, hazards and risks and responding in a timely way to that information
5. Ensure that the PCBU has and implements processes for complying with any duty or obligation the PCBU has under the WHS Act
6. Verify the provision and use of the resources and processes referred to in paragraphs 3 to 5

The Singapore Approved Code of Practice (ACoP) on Chief Executives' and Board of Directors' Workplace Safety and Health Duties (Workplace Safety and Health Council, 2022) defines company directors as the chief executive or a similar role participating in executive decisions, as well as actions concerning policy and decision-making that impact the company's overall business matters, or a significant portion thereof. Company directors are not identified based on their formal title and whether or not they are part of the company's board of directors.

In addition, the ACoP states four principles and 17 measures that company directors are expected to perform (see Table 7.4). The first principle is focused on ensuring adequate WSH governance by ensuring that there is clear WSH accountability for company directors and that WSH is considered at the highest level in the organisation. The second principle is focused on WSH culture, where company directors must set the tone and

Table 7.4 Principles and measures in the ACoP on Chief Executives' and Board of Directors' Workplace Safety and Health Duties.

Principles	Measures
Principle 1: Ensure WSH is integrated into business decisions and clarifies the roles and responsibilities of the chief executive and individual members of the board of directors in leading WSH.	1: Assign and document WSH roles and responsibilities of individual company director(s).
	2: Establish the WSH policy, standards, and strategic goals for the organisation.
Principle 2: Continuously build a strong WSH culture, set the tone and demonstrate visible leadership in embodying and communicating highly effective WSH standards.	3: Publish the organisation's WSH commitment, and review, endorse and track the organisation's WSH targets and performance regularly.
	4: Set WSH as a regular agenda item in management/board meetings.
	5: Ensure sufficient resource allocation to WSH.
	6: Facilitate direct reporting of WSH issues to the company director(s).
	7: Acquire WSH knowledge.
	8: Conduct engagements to understand processes, workers' concerns, and communicate the need to prioritise WSH.
	9: Set and demand effective WSH standards and performance from vendors and partners.
Principle 3: Ensure that WSH management systems are highly effective and reviewed regularly.	10: Ensure effectiveness of WSH management systems and maintain oversight of compliance with safe work procedures.
	11: Ensure suitable, adequate and timely risk assessment.

(Continued)

Table 7.4   (*Continued*)

Principles	Measures
	12: Recognise and reward workers' efforts toward achieving good WSH performance.
	13: Endorse immediate remedial/disciplinary actions to address workers' repeated non-compliance with safe work procedures.
Principle 4: Empower workers to actively engage in WSH.	14: Ensure processes are in place for workers to receive information on WSH risks and safe work procedures in a timely manner.
	15: Set up reporting systems, encourage proactive reporting, and ensure proper follow-up to address WSH issues.
	16: Commit resources and protected time for workers to undergo WSH training and refresher courses.
	17: Involve workers in the joint development and implementation of strategies/programmes to improve WSH.

demonstrate visible leadership through their communications, actions and behaviours. Principle 3 is focused on ensuring that the WSH management system (see Chap. 6) is effective and regularly reviewed. Principle 4 is about empowering workers to actively engage in WSH, which is also a key theme of ISO 45001, as highlighted in Chap. 6.

With the development in legislation and code of practice on company directors' WSH duties, safety leadership is no longer a good-to-have. The following is a UK case study (BBC, 2011) to illustrate the point. In 2008, Alexander Wright, a 27-year-old junior geologist employed by Cotswold Geotechnical Holdings Ltd, died in a 3.8 m deep trench collapse while collecting soil samples. The incident marked a significant moment in legal history, as it led to the first conviction under the UK Corporate Manslaughter and Corporate Homicide Act 2007. The HSE investigation revealed that the company had failed to adhere to standard health and safety regulations. The trench's sides, where Wright was working, were not supported or assessed for stability, representing a severe breach of the duty of care owed to employees.

The case went to trial in 2011, and Cotswold Geotechnical Holdings was found guilty of corporate manslaughter. The prosecution argued successfully that the company's system of work was unsafe and that the director had been grossly negligent in his duty to ensure the safety of his employees. The company was fined £385,000, to be paid over ten years. The charges against the director were dropped mainly because of his poor health arising from the stress of the prosecution process (Safety & Health Practitioner, 2011).

The "R v Cotswold Geotechnical Holdings Ltd" case underscores the profound legal and moral responsibilities that company directors hold in ensuring WSH. The key takeaways from the case in the context of safety leadership and company directors' WSH duties include:

1. The case shows that company directors need to take their WSH duties seriously. This precedence demonstrates that company directors can be taken to task for not performing their WSH duties.
2. The case highlighted that it is insufficient for companies to address safety passively; they must take a proactive stance. This involves regular risk assessments, proper employee training, adequate safety protocols, and continuous monitoring and improvement of safety measures.

3. Directors and senior management must lead by example. They are not only responsible for setting safety policies but also for instilling a culture of safety. Their commitment to workplace safety influences the entire organisation's attitude and behaviour regarding compliance and risk management.

4. Beyond the human cost, the case demonstrated the substantial financial risks companies face when neglecting WSH. The fine imposed was significant, especially given the company's size, and the reputational damage was considerable.

5. The case emphasises the need for ongoing due diligence. Companies must stay abreast of best practices in WSH and continually update policies and procedures to reflect the latest standards and regulations.

7.9.4 Project manager's workplace safety and health duties

Besides company directors, frontline managers, like project managers, are also expected to display safety leadership and prevent incidents. This section will discuss construction project managers' duties and expected attitudes, which are applicable across industries.

The Construction Regulations under the WSH Act define a project manager as someone stationed at the worksite and having overall control of the activities at the worksite. They must be competent and have gone through suitable training. In addition, the Occupier, i.e., the main contractor, must have authorised the person to act as the project manager.

Project managers are also expected to have good safety attitudes. Based on the National Curriculum, Training and Assessment Guide for the course "Manage Workplace Safety and Health in Construction Sites", project managers are expected to be effective leaders who understand the importance of safety leadership and are role models in terms of WSH. They are also expected to be self-motivated to improve WSH. Furthermore, to be an effective leader, it is important to have good communication skills. Project managers must not only be effective in communicating project scope, cost and schedule but also in emphasising the importance of WSH. They should also be able to facilitate discussions on WSH matters. In addition, project managers must be able to make sound decisions based on available WSH information and data. Project managers are expected to do everything within their means to implement the WSH management system effectively. The key word is implement. It is not just paperwork. The project manager should focus on implementing and improving the WSH management system. If the system is implemented well, it will ensure the WSH of workers during the construction stage.

Regulation 5 of the Construction Regulations states that the project manager is expected to convene site coordination meetings whenever necessary. The main aim of the site coordination meeting is to keep track of construction progress and forecast the work to be done to coordinate across the different groups of workers. Good coordination will prevent miscommunication and last-minute improvisation, reducing the risk of accidents.

The project manager must chair each site coordination meeting. Having the project manager overseeing the meeting helps ensure attendance from different main contractor personnel and subcontractors with work that requires coordination. Supervisors, engineers, and WSH personnel involved in the interacting work activities must attend the site coordination meeting. Their presence allows proper coordination to discuss issues such as locations of materials, equipment and workers, necessary communication at different points of time, and controls that should be implemented beforehand. Accidents can happen if one of the parties affects the other parties unknowingly. For example, workers might unknowingly enter a crane danger zone, and the movement of the crane can cause harm to the workers. The idea of coordination meetings came from shipyards, which require the Vessel Safety Coordination Committee meetings (VSCC). The VSCC has been used in Singapore shipyards since the 1990s to promote coordination and to prevent accidents.

Another duty of the project manager under the construction regulations is to implement the permit-to-work (PTW) system. In the PTW system, the subcontractor or the contractor in charge of a work activity would raise the PTW form. The PTW form could be digital or in hard copy, though the former is more efficient. The required information in the PTW form includes the location of work, the type of work to be conducted, safety and health risks, control measures, the people involved and their training certificates, etc. Once the form is completed, it should be submitted to the WSH Department, and an assessor or designated person will be assigned to assess whether the measures are implemented or can be implemented. Once the assessor signs off the form, it will be submitted to the project manager. After the project manager has checked and approved the permit, the permit must be displayed at the work location, and the work can progress. During the work, and depending on the type of high-risk work, the WSH Department will monitor the work activity and update the project manager periodically. The updating and reporting of PTWs can be done during the site coordination meeting. At the end of the work activity, the PTW form must be removed, and the different parties must be informed. One example is excavation under Regulation 77 of the construction regulations. If the designated person responsible for monitoring excavation work finds the work unsafe, the designated person will have to inform the project manager of the unsafe condition. The project manager can then revoke the excavation PTW, and the excavation subcontractor will have to re-apply for the PTW before continuing with work. This is how the project manager can exert control over high-risk work.

The WSH Committee Regulations also require the project manager to chair the WSH committee. The secretary of the Committee is typically the WSH officer. The WSH officer is there to advise the project manager and help run the committee. The committee will also have worker representatives and management representatives. Management representatives could be construction managers and engineers, usually executive-level employees. Worker representatives can be from the different trades and subcontractors. In Singapore, workers are frequently represented by their supervisors because of language and literacy issues, but the workers' representatives should always be there whenever suitable. The committee meets at least once a month, and the project manager must closely monitor it. The committee will also conduct monthly inspections, walk the site, and discuss the findings in the next meeting. Inspections must also be done if there are incidents so that the committee can make safety recommendations to the occupier to prevent a recurrence of the incident. The WSH committee should also promote safety and health through different activities, such as talks on safety hazards and guidelines, to help improve site safety and health.

Like other WSH duty holders, project managers can be personally liable if they do not perform their WSH duties. Any offence can result in fines and even imprisonment. These are deterrence and a reminder of the importance of WSH duties.

7.10 BP Texas City explosion case study

The BP Texas City Refinery explosion and fire (Chemical Safety Board, 2007) is one of the worst industrial accidents in recent history. The accident happened on 23 March 2005 at 1:20 pm (GMT-5). The explosion and resulting fire killed 15 people and injured another 180. The financial losses were estimated to be at least US$1.5 billion. The explosion occurred during the start-up of an isomerisation (ISOM) unit when a raffinate splitter tower was overfilled. The overfilling led to the opening of pressure relief devices. Still, it resulted in a flammable liquid geyser (a fountain of high-pressure jets of fluid and gas that shoots into the air) from a blowdown stack that did not have a flare (a safety device which can burn off gas or liquid released through the blowdown). This resulted in the formation of a flammable vapour cloud that was ignited by the idling engine of a pickup truck parked near the blowdown stack.

The investigations conducted after the accident revealed severe organisational and safety culture problems in BP Texas City Refinery. Some of them are provided below:

1. The BP Board of Directors did not provide an effective oversight of BP's safety culture and major accident prevention programmes. The Board did not have a member responsible for assessing and verifying the performance of BP's major accident hazard prevention programmes.

    Note: This shows that safety leadership was inadequate. An informed culture where senior management knew about key safety issues was not present.

2. Numerous surveys, studies and audits identified deep-seated safety problems at BP Texas City, but the response of BP managers at all levels was typically "too little, too late".

    Note: This is a sign of a lack of proactive actions by managers and leaders, i.e., a safety leadership issue.

3. Cost-cutting, failure to invest, and production pressures from BP Group executive managers impaired process safety performance at BP Texas City.

Note: This reflected how cost-cutting was valued over process safety — a sign of poor safety culture and a lack of safety leadership.

4. A "check the box" mentality was prevalent at BP Texas City, where personnel completed paperwork and checked off on safety policy and procedural requirements, even when those requirements had not been met.

   Note: Routine violation was common, and such behaviour is again a reflection of unsafe shared values and beliefs. "Check the box" is the same phenomenon as the "paper exercise" highlighted in Chap. 4. These phenomena are reflections of surface compliance (Hu *et al.*, 2020), which is compliance in form but not in intent. In contrast, deep compliance is focused on compliance with the rule's intent (Hu *et al.*, 2020).

5. BP Texas City lacked a reporting and learning culture. Personnel were not encouraged to report safety problems, and some feared retaliation for doing so. Therefore, lessons from incidents and near-misses were generally not captured nor acted upon. Important relevant safety lessons from a British government investigation of incidents at BP's Grangemouth, Scotland refinery were also not incorporated at BP Texas City.

   Note: This is aligned with Reason's (1997) model of safety culture, where reporting and learning culture are critical components.

6. Reliance on the low personal injury rate at BP Texas City as a safety indicator failed to provide a true picture of the processes' safety performance and the health of the safety culture.

   Note: This is a poor choice of WSH indicator, which reflects an inadequate WSH management system. There is a lack of suitable indicators to reflect the plant's process safety.

7. Safety campaigns, goals and rewards focused on improving personal safety metrics and worker behaviours rather than

on process safety and WSH management systems. While compliance with many safety policies and procedures was deficient at all levels of the refinery, BP Texas City managers did not lead by example regarding safety.

Note: This is a major management system issue similar to the previous point.

The above findings showed that an inadequate WSH management system, poor safety culture, and inadequate safety leadership were intertwined, and they contributed to the BP Texas City explosion. These underlying factors are almost always found in major accidents, albeit with variances in the details. For example, the investigation into the flooding of the Bishan train tunnel, which led to a 20-hour train service disruption in Singapore on 7 October 2017 (Kuek, 2017), showed that even though the flooding was triggered by heavy rain, the incident was a result of a poorly implemented maintenance system (including falsified maintenance records) contributed by "deep-seated cultural issues". Then Transport Minister Khaw Boon Wan said, "[A] positive corporate culture starts from the top." (Tan, 2017) At the same time, he emphasised, "[G]rowing the right culture is the responsibility of everyone — from the top leadership down to the workers." This is true because culture is the product of shared values and beliefs. Everyone influences culture, including safety culture, but top management is expected to be directly accountable for culture.

7.11 Learning disabilities and safety culture

Reason (1997) highlighted the importance of learning culture as a sub-culture of an informed culture and safety culture. At the same time, WSH management standards such as ISO 45001:2018 (see Chap. 6) are based on the Plan–Do–Check–Act (PDCA) cycle, which emphasises the importance of continual improvement and learning. Senge (2006, p. 4) highlighted, "[T]he organisations that will truly excel in the future will be the organizations that discover how to tap people's commitment and capacity to learn at all levels in an organization." An organisation that "truly excels" must have a strong safety culture to sustain WSH performance across time. However, many organisations have learning disabilities that can lead to poor learning and safety culture. Senge (2006) highlighted seven learning disabilities:

1.   **I am my position**

Over time, individuals become attached to their positions, identifications and roles, and this leads to a "silo mentality", where they are no longer concerned with or become unable to see how their actions affect the other parts of their organisation and vice versa. When procurement executives are asked to consider the safety and health of workers, they cannot see why they need to do so because, to them, it is the WSH officer's (WSHO) job to do so and not theirs. The procurement executive did not realise that the procurement department introduces many workplace hazards when unsafe equipment or incompatible materials are purchased. A myopic view of roles is common, and people do this to protect themselves from overwork and to keep roles and responsibilities clear and systematic. There is always a concern that once they help "others" to do their work, their job scope will become enlarged and unmanageable. Such a mentality focuses on what people do and not the purpose of the organisation that they are in. This learning disability limits the possible actions that organisations can take when facing complexity, and the full potential of an organisation can never be fully utilised.

2.   **The enemy is out there**

I had observed parents who comfort their crying child who had just hit the wall by slapping the wall and saying, "Ok, I'm punishing the wall for hitting you on the head. Don't cry anymore..." Many people have the tendency to blame others for the undesired situation they are in, but a friend of mine was extraordinary. We were trying to fix something on the ceiling, and the ladder we had was too short, so I said, "The ladder is too short." The friend replied in a philosophical tone, "No, I am too short." Such a mentality is remarkable. Both of us were not wrong, but his response showed a fundamental shift in mindset. He looks for his contribution to the problem we are facing, and such a mentality allows organisations to look inward for ways to deal with complex problems and achieve the desired outcomes.

An example provided by Wright (Meadows and Wright, 2008) further illustrates the learning disability of "the enemy is out there". She first showed her class how a Slinky bounced up and down when it hung from her hand. She asked her students, "What made the Slinky bounce up and down like that?" The students said her hand caused the Slinky to bounce. When Wright hung a small box from her hand, the box did not bounce like the Slinky. Both the Slinky and the box were placed in the same situation (hung from her hand), but the responses were drastically different. The internal structures of the Slinky and the box led to the differences in behaviour. Internal structure refers to the way a system's different elements or components are connected, e.g., the way materials are joined to form the Slinky and the box.

Similarly, a manager can complain that the client is the one who is causing all the WSH problems by not putting enough emphasis on WSH. This could lead to the manager's failure to identify ways that he and his team can better manage their WSH management system and culture.

3.   The illusion of taking charge

Being proactive in doing "something" might actually be reactiveness in disguise. When a manager starts to be proactive and take charge when dealing with safety culture problems, he might take an aggressive stand against an "enemy out there". This would worsen the problem of "the enemy is out there". For example, the manager might think that it is the subcontractors' fault for not complying with WSH rules given to them, and he did not consider the need to review their contractor selection process and other related processes to assess how they could have selected better contractors to assist in their projects.

4.   The fixation on events

As explained in earlier chapters, events are episodes of significance. When managers are fixated on events, it means that they do not look for the underlying structure or factors that influence the event occurrence. When a project has an accident, it is easy to blame it on the worker who caused the accident. However, as highlighted in Chap. 3, it is more important to understand the management system's inadequacies, safety culture, and leadership issues that led to the accident. Focusing on events leads to event-level explanations and solutions, such as firing the WSHO or blaming individual workers, which does not help the organisation to learn and improve.

5.   The parable of the boiled frog

The parable of the boiled frog is about our insensitivity to gradual changes in our environment. The parable says that if a frog is placed in a pot of water and the temperature of the water is raised slowly, the frog will do nothing to escape the impending danger. It might start to feel uncomfortable at some point, but by the time the frog notices the danger of the rising temperature, it is too late, and it gets boiled alive. This sounds very much like the climate change challenges that the world is experiencing. From a WSH point of view, managers need to set up a monitoring system to continuously assess the changes in the organisation and its environment. One of the common mistakes managers make is to belittle possible WSH issues raised by team members or stakeholders only to realise later on that those issues lead to WSH incidents (as will be observed in Chap. 10). Safety culture changes gradually over time and managers need to monitor safety culture and proactively maintain or improve it. Consistent and deliberate considerations can help prevent the organisation from becoming a frog in boiling water.

6.   The delusion of learning from experience

We learn best from experience, but that is only true if we can see the consequences of our actions. If there is a significant delay between action and consequence, or if the consequences of our actions are not known to us, then it is not easy to learn from such experience. For example, a manager might not have been communicating with his team frequently, and he will not know that his lack of communication is slowly causing him to be alienated from his team. This further deteriorates communication between the manager and his team, leading to a vicious cycle where lack of communication leads to even less communication in the team. The manager might think that by cutting down on meetings, he has successfully improved his team's productivity, but the lack of meetings has had unintended effects on his team. However, the manager will only experience the negative consequences after some time, which becomes a barrier to him realising his mistakes.

7.   The myth of the management team

The management team is supposed to be a team of individuals who work together to bring success. However, the reality is that most teams in workplaces are not performing as a team. Instead, individuals might be more concerned with individual key performance indicators (KPIs) and office politics and turfs. Many teams do not communicate well and break down under pressure arising from complexity. Dysfunctional teams cannot decipher complexity and derive solutions to handle the complexity of the work, which can result in WSH incidents. Successful management teams have individuals willing to let go of their egos and territories and humbly explore complex problems with their team members. They are open-minded and willing to listen to others. Such teams are hard to come by but extremely valuable and powerful when handling complexities.

These seven learning disabilities lead to organisations being unable to manage hazards and risks arising from work, causing injuries and ill health. On the other hand, organisations "where people continually expand their capacity to create the results they truly desire, where new and expansive patterns of thinking are nurtured, where collective aspiration is set free, and where people are continually learning how to learn together" (Senge, 2006, pp. 3–4) are learning organisations that can excel and achieve excellent WSH performance. Senge (2006) further postulated that there are five fundamental disciplines critical to the development of a learning organisation. This ensemble of five disciplines will be discussed in the next chapter.

7.12   Conclusions

Safety culture is about an organisation's shared values and beliefs that influence safety-related behaviours. There are various safety culture models, and some are discussed in this chapter. Regardless of the definitions and models, the key thing to note is that safety culture influences, albeit not in a deterministic manner, human behaviours while implementing WSH procedures and policies. If there is no buy-in from the people implementing the WSH management system, the system will never work. When dealing with people, we need to use humanistic language and be in sync with emotions. Despite the importance of legislation, liabilities and statistics, if there is an over-focus on them, it is not possible to get WSH messages across. Leaders and managers have to demonstrate genuine care for the workers to let the workforce know that everybody needs to use and improve the WSH management system. Thus, safety leadership is intertwined with safety culture, and the key to a positive culture rests with senior management.

Safety leadership is essential to WSH management, and company directors must manage WSH proactively and demonstrate their commitments. Prosecution cases show that company directors can face severe consequences if they do not perform their WSH duties. Furthermore, the BP Texas City case study shows that major accidents tend to occur because of the inter-relationships among the WSH management system, safety leadership, and safety culture factors.

A learning organisation, as proposed by Senge (2006), is aligned with a positive safety culture. The seven learning disabilities highlighted by Senge (2006) are some of the possible barriers to achieving a learning culture, which is essential for a positive safety culture.

Review questions

1. Explain Reason's model of safety culture and contrast it with the HSE Safety Culture questionnaire.
2. Why is top management so important in improving safety culture?
3. What can top managers do to demonstrate their commitment to WSH?
4. Explain the difference between safety culture and safety climate.
5. Explain the difference between construction safety climate and DfS climate.
6. Differentiate transactional and transformational leadership.
7. Describe the different leadership approaches in the Situational Leadership® II model.
8. List some of the safety culture problems at BP Texas City Refinery prior to the accident in 2005.
9. Using the four principles of safety leadership stated in the ACoP for company directors' WSH duties, discuss how senior management could have prevented major accidents in the context of the BP Texas City Refinery fire and explosion.

10. Explain how the seven learning disabilities described in (Senge, 2006) contradict the formation of the positive safety culture described in (Reason, 1997).

References

Barling, J., Loughlin, C., and Kelloway, E. K. (2002). Development and test of a model linking safety-specific transformational leadership and occupational safety. *Journal of Applied Psychology*, **87**(3), 488.

Bass, B. M. and Avolio, B. J. (1990). *Transformational leadership development: Manual for the multifactor leadership questionnaire.* Consulting Psychologists Press.

BBC. (2011). R v Cotswold Geotechnical Holdings Ltd [2011] EWCA Crim 1337 | Croner-i. 17 February. https://app.croneri.co.uk/law-and-guidance/case-reports/r-v-cotswold-geotechnical-holdings-ltd-2011-ewca-crim-1337

Blanchard, K. H. (1985). *SLII®: A situational approach to managing people.* Escondido, CA: Blanchard Training and Development.

Blanchard, K., Zigarmi, P., and Zigarmi, D. (2013). *Leadership and the one minute manager: Increasing effectiveness through Situational Leadership® II.* New York: William Morrow.

Chemical Safety Board. (2007). Investigation report — Refinery explosion and fire (BP Texas). http://www.csb.gov/ (accessed: December 2008).

Chen, H., Li, H., and Goh, Y. M. (2021). A review of construction safety climate: Definitions, factors, relationship with safety behavior and research agenda. *Safety Science*, **142**, 105391.

Clarke, S. (2013). Safety leadership: A meta-analytic review of transformational and transactional leadership styles as antecedents of safety behaviours. *Journal of Occupational and Organizational Psychology*, **86**(1), 22–49.

Downton, J. V. (1973). *Rebel leadership: Commitment and charisma in the revolutionary process.* New York: Free Press.

Duhigg, C. (2014). *The power of habit: Why we do what we do in life and business.* New York: Random House.

Energy Institute. (n.d.). Hearts and minds. http://www.eimicrosites.org/heartsandminds/ (accessed: 8 October 2015).

Guldenmund, F. W. (2000). The nature of safety culture: A review of theory and research. *Safety Science*, **34**(1–3), 215–257.

Guldenmund, F. W. (2007). The use of questionnaires in safety culture research — an evaluation. *Safety Science*, **45**(6), 723–743.

Health and Safety Executive. (n.d.). Common topic 4: Safety culture. www.hse.gov.uk/humanfactors/topics/common4.pdf   (accessed: 2 September 2020).

Hofmann, D. A., Burke, M. J., and Zohar, D. (2017). 100 years of occupational safety research: From basic protections and work analysis to a multilevel view of workplace safety and risk. *Journal of Applied Psychology*, **102**(3), 375–388.

Hu, X., Yeo, G., and Griffin, M. (2020). More to safety compliance than meets the eye: Differentiating deep compliance from surface compliance. *Safety Science*, **130**, 104852.

Huang, Y.-H., Verma, S. K., Chang, W.-R., Courtney, T. K., Lombardi, D. A., Brennan, M. J., *et al.* (2012). Supervisor vs. employee safety perceptions and association with future injury in US limited-service restaurant workers. *Accident Analysis & Prevention*, **47**, 45–51.

Kelloway, E. K., Mullen, J., and Francis, L. (2006). Divergent effects of transformational and passive leadership on employee safety. *Journal of Occupational Health Psychology*, **11**(1), 76.

Kouzes, J. M. and Posner, B. Z. (2002). *The leadership challenge*. San Francisco: Jossey-Bass.

Kuek, D. (2017). In full: SMRT CEO Desmond Kuek on "deep-seated cultural issues" behind history of service disruptions. *Today*. https://www.todayonline.com/singapore/smrt-ceo-desmond-kueks-statement-response-oct-7-tunnel-flooding-nsl (accessed: 6 April 2020).

Lewin, K., Lippitt, R., and White, R. K. (1939). Patterns of aggressive behavior in experimentally created "social climates". *The Journal of Social Psychology*, **10**(2), 269–299.

Lim, S. H. M. and Goh, Y. M. (2024). Development and validation of the Design-for-Safety (DfS) climate measurement tool. *Journal of Risk Research*.

Meadows, D. H. and Wright, D. (2008). *Thinking in systems: A primer*. White River Junction, Vt: Chelsea Green Pub.

Northouse, P. G. (2016). *Leadership: Theory and practice*. Thousand Oaks, CA: Sage Publications Inc.

Reason, J. (1997). *Managing the risks of organizational accidents*. Aldershot: Ashgate.

Roth, M. (2012). "Habitual excellence": The workplace according to Paul O'Neill. http://www.post-gazette.com/business/business-news/2012/05/13/Habitual-excellence-The-workplace-according-to-Paul-O-Neill/stories/201205130249 (accessed: 8 October 2015).

Safe Work Australia. (n.d.). The health and safety duty of an officer. https://www.safeworkaustralia.gov.au/system/files/documents/1812/officer-duty-interpretive-guide.pdf (accessed: 16 October 2023).

Safety & Health Practitioner. (2011). First corporate manslaughter conviction delivers £385,000 penalty. *Safety & Health Practitioner*, 17 February. https://www.shponline.co.uk/legislation-and-guidance/first-corporate-manslaughter-conviction-delivers-385-000-penalty/

Senge, P. (2006). *The fifth discipline — The art & practice of the learning organisation*. New South Wales: Random House Australia.

Tan, C. (2017). Parliament: Right culture starts from top, says Khaw Boon Wan. *The Straits Times.* https://www.straitstimes.com/singapore/transport/parliament-right-culture-starts-from-top-says-khaw. (accessed: 6 April 2020).

Uttal, B. (1983). The corporate culture vultures. *Fortune*, 17 October, 66–72.

Workplace Safety and Health Council (2015). Apply for CultureSAFE. http://www.wshc.gov.sg/ (accessed: 8 October 2015).

Workplace Safety and Health Council. (2022). Code of Practice on Chief Executives' and Board of Directors' WSH Duties. https://www.tal.sg/wshc/resources/publications/codes-of-practice/code-of-practice-on-chief-executives-and-board-of-directors-wsh-duties (accessed: 13 October 2023).

Zohar, D. (1980). Safety climate in industrial organizations: Theoretical and applied implications. *Journal of Applied Psychology*, **65**(1), 96–102.

Zohar, D. (2010). Thirty years of safety climate research: Reflections and future directions. *Accident Analysis & Prevention*, **42**(5), 1517–1522.

Zohar, D. and Luria, G. (2005). A multilevel model of safety climate: Cross-level relationships between organization and group-level climates. *Journal of Applied Psychology*, **90**(4), 616–628.

CHAPTER 8

Improving safety culture

The Event Causation Technique

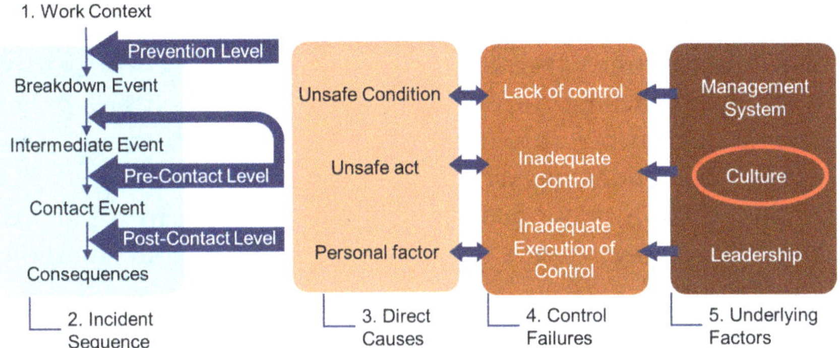

As established in the earlier chapters, safety culture interacts with management systems to influence behavioural norms and, hence, incident occurrence. To improve safety culture, the organisation needs to undergo fundamental changes, which can be implemented based on organisational change theories and concepts.

8.1 Introduction

Organisational change refers to fundamental alterations in an organisation's core elements, such as vision, mission, strategies, culture, structure, competencies, critical management system processes, and personnel. In the context of WSH management, organisational change includes safety culture improvement initiatives, in which managers, project managers, engineers, and workplace safety and health (WSH) managers and officers play a crucial role. This chapter offers an overview of how organisational change management concepts can be leveraged to enhance safety culture within an organisation.

It was highlighted in Chap. 6 that management of change (MOC) is an important process in a WSH management system. This process uses a risk-based approach to minimise the risk of operational changes leading to hazards that cause accidents or ill health. Instead of MOC, this chapter will discuss change at the organisational level, specifically, improvement in safety culture. In the context of WSH, the motivation for safety culture improvement frequently reacts to external pressures, especially enforcement from a regulator or client after a major accident. Other possible drivers for a safety culture initiative include new leadership with a focus on WSH, a merger with a company with a stronger safety culture, pressure from unions, new WSH management standards or legislation, and changing industry norms.

After a major accident, an organisation's survival and future success depend on its ability to improve its safety culture and WSH management system in response to internal and external pressures. If the organisation cannot demonstrate its ability to ensure workers' safety and health, they may be put out of business. Furthermore, managers and leaders may be removed from their positions. Thus, organisations must proactively improve their safety culture and WSH management system and not wait for a major incident to provide the impetus. However, most organisational change initiatives fail, so we must understand how to design and implement successful change.

8.2 Improving safety culture

Using the definition of safety culture introduced in the last chapter, three possible approaches to improving safety culture can be identified (see Fig. 8.1).

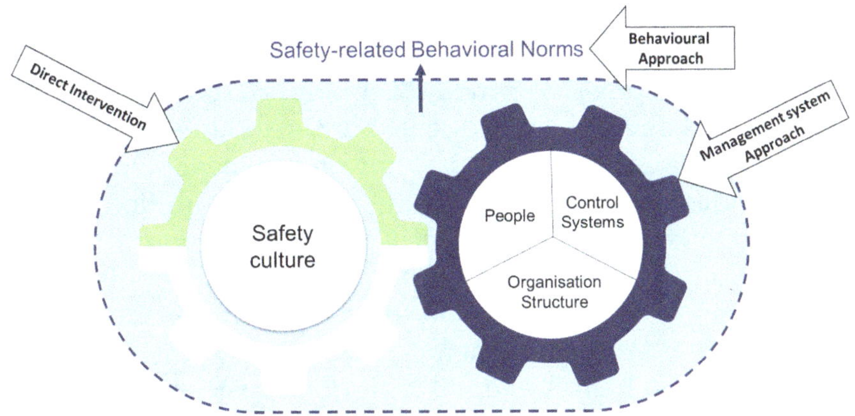

Fig. 8.1   Approaches to improve safety culture.

First, a direct intervention approach focuses on the middle layer of safety culture, i.e., espoused values and attitudes. In this approach, management aims to express, communicate, and diffuse desired safety beliefs and values throughout the organisation so that all employees share the vision of a safe and healthy workplace. This can be done through facilitated focus group discussions, interviews, and safety climate surveys. Focus group discussions and interviews will produce qualitative cases or stories that provide anecdotal evidence of shared values and beliefs. These stories can be very powerful in further establishing shared values and beliefs. The effectiveness of directly discussing safety culture depends on the quality of communication and facilitation. A safety climate survey provides a quantitative source of information indicative of the shared values and beliefs that can be used to triangulate with the information obtained from focus group discussions and interviews. The direct intervention approach assumes that safety culture can be verbalised, discussed, and modified through discussions.

Second, safety culture can be improved by focusing on the WSH management system. This approach is an outside-in approach. Since management standards (e.g., ISO 45001 and

ISO 14001) (see Chap. 6) are essentially good practices adopted from organisations with positive safety culture, it is assumed that organisations that model themselves after these good practices will develop a good safety culture. This is focused on the outer layer of safety culture, but since a management system includes training, communication and consultation, there will be overlaps with the direct intervention approach. The assumption is that when the management system is well-designed and implemented, and there are consistent checks to improve the WSH management system, the process will cultivate values and beliefs aligned with good WSH management principles.

The last approach is a behavioural approach, which focuses on the behaviours of members of the organisations. Behaviours arise because of intentions, motivations and consequences of the behaviours. They are more observable than values and beliefs and can be used as safety culture indicators. A quantitative approach can be used to measure behaviours. Tracking and trending the behaviours across time will provide managers with data, which they can use as a leading indicator of the effectiveness of WSH management. During behavioural observations, the observer will be trained to provide positive reinforcement or feedback to motivate the person being observed to work safely. Such an approach is known as the behaviour-based safety (BBS) approach, and there are many variations to how BBS is implemented. New technologies like computer vision and wearables can also be incorporated to facilitate the implementation of BBS. BBS assumes that changing behaviour can improve safety culture (shared values and beliefs).

All three approaches are overlapping, viable and interdependent. An effective safety culture improvement initiative will involve using all three approaches in different stages of the programme.

8.3 Cultivating safety culture through disciplines of learning organisation

As highlighted in Chap. 7, Senge (2006) emphasised the importance of a learning organisation in coping with changes and complexities. Like other organisational changes, safety culture improvement initiatives are always complex because of the uncertainty, ambiguity, and complex interdependencies between people and tasks. According to Senge (2006), systems thinking helps an organisation understand complexity, but a learning organisation needs to practise four other disciplines that can be grouped into two groups. The first group is "aspirations", which includes the disciplines of personal mastery and shared vision. The second group is "reflective conversation", which includes mental models and team learning and dialogue. All four disciplines are related to the direct intervention approach, particularly shared vision, team learning and dialogue. The following gives an overview of the disciplines.

Senge (2006) considers systems thinking the cornerstone of a learning organisation. Systems thinking is a way of seeing systems (including WSH management systems) holistically. A systems thinker would strive to understand the underlying patterns, structures and factors influencing the occurrence of WSH incidents so that more fundamental solutions can be derived to achieve sustained positive WSH performance even in complex and challenging environments. We have introduced systems thinking tools such as causal loop diagrams (Chap. 1) and system archetypes (Chap. 3); these are useful tools to evaluate and improve safety culture, but the most important contribution of systems thinking to safety culture is the proactive mindset of seeing how our actions contribute to systemic problems. In seeking to understand the system, we need to look at how different system elements, including ourselves, interact to produce the behaviours we observe. This holistic and inward-looking approach changes an organisation's fundamental beliefs and values, i.e., organisational and safety culture will be changed.

Personal mastery (Senge, 2006, p. 7) "is the discipline of continually clarifying and deepening our personal vision, of focusing our energies, of developing patience, and of seeing reality objectively". An organisation is made up of individuals, and a learning organisation can only be formed if the individuals in the organisation are truly committed to learning. In our context, individuals must learn how to make the workplace safe and healthy, and the organisation must encourage and facilitate this commitment. Personal mastery is essentially about employees learning to create and sustain the "creative tension" or resisting the pressure to give up their vision when facing the current reality. This is related to the archetype of "eroding or shifting goals" (Chap. 3), where individuals and their organisation must "hold the vision" even when the current reality is way below the vision.

Mental models (Senge, 2006, p. 8) "are deeply ingrained assumptions, generalizations, or even pictures or images that influence how we understand the world and take action". Based on our definition of safety culture, shared mental models of WSH are essentially an organisation's safety culture. In the context of WSH, the discipline of mental models is about looking inwards to understand our basic assumptions about WSH (e.g., our assumptions about how WSH incidents happen) and evaluate them. This can be the basis for "learningful" conversations that balance inquiry and advocacy.

The discipline of building shared WSH vision is about safety leadership. Good WSH leaders inspire and sustain a vision of an injury- and illness-free organisation. They rally the people to be committed to WSH goals, values and mission. The individuals must truly share the vision, and, as discussed in Chap. 7, different leadership approaches can be adopted to build the shared vision. However, it must be noted that leadership includes more than just the top management. Anyone in the organisation can be a WSH leader by being committed to the WSH-related activities and leading through their actions. A shared vision is critical to a safety culture improvement initiative because it helps the organisation focus its energy and resources and establishes a reference point for everyone. Safety culture, defined as shared values and beliefs about WSH, is directly related to a shared vision. Suppose the employees of an organisation are truly committed to the same vision of zero injuries and illnesses. In that case, they will naturally value WSH and believe they must do more to achieve the shared vision. Building a shared vision involves individuals having their personal vision through the discipline of personal mastery.

Building a shared vision and improving an organisation's safety culture requires good team learning and dialogue. Team learning and dialogue is about building high-performance teams that can learn as a team, and everyone can tap into each other's areas of competence to solve problems and achieve team goals. Senge (2006, p. 9) indicated that team learning starts with team dialogue or the ability to "suspend assumptions and enter into a genuine 'thinking together'". During an effective team dialogue, team members explore a complex issue, e.g., WSH-related issues, from different angles to gain clarity and fundamentally improve the system. This can only arise if the team has a shared vision and can let go of their egos and personal gains to focus on the "win" at the system level. Effective team dialogue results in team learning, and if there are enough teams doing this in the organisation, organisational learning is achieved. Team learning is also about identifying possible team learning dysfunctions and actively preventing them from happening. In WSH management, the basic unit is usually a team. Risk assessment (RA), incident investigation, and many other WSH management activities are frequently conducted at team level. If an RA team cannot have open and meaningful discussions about hazards, it is a clear sign that the team, and perhaps the organisation, have a weak safety culture. Therefore, safety culture is closely related to team learning.

During direct interventions on safety culture, where dialogues are being conducted to discuss safety culture, participants of the dialogue sessions must suspend their assumptions and regard one another as colleagues. In addition, the facilitator must be able to hold the context of the dialogue (Senge, 2006). After the dialogues, discussions must be conducted so that participants advocate possible visions and approaches the team must deliberate and adopt. This process of dialogue (open exploration) and discussion (advocacy and decision-making) is an important communication and consultation process in the direct intervention approach.

As can be observed, the concept of a learning organisation is closely related to safety culture, and organisations trying to improve safety culture can use the five disciplines proposed by Senge (2006) as a guiding framework for their interventions.

8.4 Change management models

8.4.1 Lewin's change model

One of the well-known change models is Lewin's change model (Cummings *et al.*, 2015), which states that change goes from unfreezing to moving and then refreezing. Unfreezing involves the removal of resistance to change through understanding the reason for resistance and providing information and education to the relevant stakeholders. Moving refers to the process of changing behaviours, attitudes and values by establishing new policies, structures and processes. Refreezing creates a new equilibrium by reinforcing the changes and ensuring sustainability. Before the unfreezing stage, the organisation must identify the need for change and the problems that can arise if the change is not implemented. Throughout the change process, data must be collected about the current situation, the change process, and the situation after the change process has been completed. In addition, the team overseeing the change will have to develop frequent communication and coordination opportunities for stakeholders to give feedback and share their challenges and concerns.

When applied to WSH culture improvement, unfreezing refers to the process of understanding and removing the reasons holding back the culture improvement process. Moving refers to the process of establishing safety attitudes and values. Refreezing refers to the process of maintaining the established safety culture. This simple model helps managers plan and structure their safety culture improvement plan.

8.4.2 Rider-Elephant-Path model

Heath and Heath (2010) gave an interesting analogy of the change process. According to them, each person has an emotional elephant and a rational rider. The rider riding the elephant is objective and rational but may not be able to control the big elephant, which is focused on the short term and is guided by feelings. At the same time, if you want to influence the behaviour of the rider and his elephant, you will have to shape the path that they are trudging on. Based on this analogy, Heath and Heath (2010) suggested nine key change management actions — three for each of the components in the analogy (the rider, the elephant, and the path). Even though the following discusses each component separately, all three components interact with each other, and it is difficult to distinguish the effects of each component clearly. In addition, not all nine change management actions are relevant in all situations. Change leaders need to select a suitable combination of actions to encourage change.

8.4.2.1 Direct the rider

The Rider is a "thinker and planner" but tends to "overthink" (decision or analysis paralysis) and over-focus on problems. To direct the rider towards the desired behaviour, a leader should find "bright spots" (positive examples) for followers to follow, clearly script the critical actions, and make the desired outcome clear.

Finding the bright spot is about looking for "insiders" who have used good or best practices to achieve desired results or top performance. An important note is that the manager should look for practice, not just knowledge. For example, many contractors know it is important to avoid working at height (knowledge), but few practise work methods that minimise working at height consciously. Furthermore, it is important to emphasise bright spots that arose from the "locals" (of a project or company) so that the "not invented here" problem will not arise, and workers will be more receptive to the good practice. It is also interesting to note that many good or best practices are easily implemented. Once the bright spots are identified, leaders must thoroughly investigate why the bright spots are different from the rest. The success factors must be identified so that it is possible to clone and reproduce as many bright spots as possible. In this way, the norm in the group changes. Cloning involves getting employees to shadow identified bright spots and getting bright spots to conduct sharing sessions.

Finding a bright spot is essentially a question of "What's working, and how can we do more of it?" It is solution-focused and not problem-focused. It is about scaling up solutions.

Since the rider can easily overthink, change leaders need to narrow down the possible options, script the critical moves, and direct workers' energy towards the desired behaviour. Too many options delay change, and employees may choose the status quo (i.e., refuse to change) if it is too difficult to decide. Thus, resistance to change may be due to a need for more clarity on the desired behaviour. For individuals to implement the change, there must be clear guidelines on the specific desired behaviour. Because it is impossible to script every move, managers must focus on the critical moves. For example, a company might want workers to be safe — but what is "being safe"? This is where organisations, through consultation with workers and supervisors, must define the specific safe behaviours for each type of trade and task.

Safe behaviour can be defined through safe work procedures (SWPs), discussed in Chap. 4. Employers and principals must conduct RAs and derive SWPs based on the RAs. The SWPs must document the specific risk controls that need to be implemented. Managers, supervisors and workers are expected to implement the risk controls, but if the SWPs are not clearly written and communicated, employees may not be able to implement the SWPs on the ground. The instructions for implementing the SWPs need to be written in a form that is easy for employees to understand.

Concurrently, organisations must ensure that the goals or desired outcomes are crystal clear so the rider can focus their plans and decisions towards the goals. At the same time, having clear and desirable goals will also motivate the elephant.

8.4.2.2 Motivate the elephant

The elephant is the rider's biggest challenge. To motivate the elephant, find the feelings that influence people's behaviour (e.g., the joy of being with one's family), shrink the change into comfortable bite-sized or baby steps, and grow the people to feel a sense of identity and believe they can change.

Heath and Heath (2010) cited Kotter and Cohen (2002), who observed that the change process in most successful change efforts is not Analyse–Think–Change, but See–Feel–Change. In the Analyse–Think–Change model, a person analyses the situation, thinks of solutions and then changes to implement the solutions. The reality is that when we see a situation or something that grabs our attention, our emotions are stirred, and they influence our decisions and responses towards the situation. Thus, if the feeling that arises is not aligned with the intended change, change will not occur. For example, suppose a campaign on "safe hands" is being implemented at a factory, even if many workers are invited to attend talks on how to protect their hands because of the cognitive bias of overconfidence (discussed in Chap. 4). In that case, there is a tendency for the workers to feel that the likelihood of themselves being the victim of a hand-related accident is very small. In such a situation, the workers tend to ignore the safety measures promoted in the campaign. The campaign did not motivate the elephant because the messages did not stir any feelings nor help the workers relate to the situation. To motivate the elephant, it is more impactful to link the desired change or behaviours to positive emotions like happiness, camaraderie among workers, and recognition rather than negative emotions like shame, the fear of being punished, and the fear of losing one's family.

"Shrink[ing] the change" refers to cutting up a change into bite-size chunks so that it is not too daunting to implement. The elephant "hates doing things with no immediate payoff". Therefore, the first step, or first five minutes is critical to evoke change. If the employee is willing to start implementing the intended change, then there is a chance for the momentum to pick up so that the change might be implemented. The first task for effecting change should be simple and meaningful, and employees should be recognised for their initial effort.

Another approach to motivating the elephant is to grow the people through inspiration so that they become bigger than the problem. The crux is the identity that people adopt. If each employee sees themself as a safety champion or the workplace as an extension of the family or friends, then helping to spot hazards and unsafe behaviour becomes natural. Thus, the key challenge for change leaders is making the intended change a matter of identity and not merely a detached cost-benefit analysis. The key test is this, "Would an employee (who is being asked to change) aspire to be the kind of person who would make this change?" If yes, the elephant can be more easily motivated to initiate the change. If the answer is no, then there is a need to review whether the change is suitable or if the perceived identity needs to be redesigned. The change leader must be mindful of the following identity questions when considering their change initiative — "Who am I? What kind of situation is this? What would someone like me do in this situation?"

Another area that the elephant might need help with is that change, especially concerning organisational issues like safety culture, is a long and challenging path. How can one continue to motivate the elephant? Employees must be convinced that though there will be failures along the way, these expected failures do not end the mission. This mentality is an important aspect of the growth mindset, which praises effort rather than natural skill and believes in the ability of people to grow and adopt new skills along the way. Thus, a learning culture (as highlighted in Chap. 7), which includes the belief that people can change and improve, is critical to building a safety culture. When incidents and accidents occur as a company embarks on a change journey to improve its safety culture and safety performance, employees' morale may be affected, especially those directly involved in the incidents or implementation of change. This can derail the whole change effort. Thus, employees must be prepared to face failures and learn from them.

8.4.2.3 Shape the path

To shape the path, organisations must tweak the environment so that it is supportive of the desired behaviour, build habits through action triggers, checklists or other aids, and make use of the herd effect to spread the behaviour that the change process is hoping to build.

There is a natural tendency to blame individuals for their errors, but as discussed in earlier chapters (e.g., Chap. 2), the errors or unsafe behaviours were frequently induced by the environment that they were in. For example, in the past, some Automated Teller Machines (ATMs) were designed to deliver the withdrawn cash before returning the card, and this caused many people to forget to retrieve their ATM cards. A simple change in sequence, where the ATM card is returned before the money is given to the customer, resulted in a significant drop in the chances of customers forgetting to retrieve their ATM cards. The idea of tweaking the environment overlaps with the field of human factors and ergonomics, which design the environment and equipment to suit human characteristics, alleviate human limitations, and minimise errors.

Ensuring a neat work environment is another simple but effective approach to improving safety behaviour. For example, suppose a construction site has poor housekeeping, where objects are placed near the edges of openings or buildings, and there are slipping hazards. In that case, it contravenes Regulations 24 and 26 of the WSH (Construction) Regulations 2007 in Singapore. When workers work in an environment where WSH contraventions are the norm, they can easily form unsafe attitudes, and there is a higher tendency to violate other WSH rules and standard operating procedures. On the other hand, if workers work in an environment where house-keeping is consistently practised and equipment and materials are kept in the correct locations, the perceived norm would be that WSH rules and procedures are important and adhered to. This will lead to a change in the behaviour of individuals placed in this environment.

Shaping the path is also about creating automatic habits. To trigger these habits or routines, change leaders can develop action triggers such as posters, signage, and hand signals, which can be incorporated into training. During training, workers can be trained on abbreviations like STOP (Stop, Think, Observe and Plan). Complementary posters can be placed in the work-place to trigger the routines taught in training, which include pausing work to think about the possible hazards and controls in the current task and location and to observe the task and location so that a plan can be put in place to ensure that the work can be conducted safely. These action triggers must be easy to understand, and the actions required must be carefully scripted, as discussed earlier.

Another commonly used trigger is a checklist. A crane operator must conduct a pre-operation check according to a checklist. The checklist is a simple intervention, but it helps ensure the check is thorough and prevents over-confidence. However, to ensure that the checklist is used properly, the operator must understand its importance and supervisors must inspect the completed checklists every day and ask probing questions regarding the checklist. If the crane operators feel (i.e., the elephant in the operators) that the checklists are for novices, they will not be used even if they are useful. One way to minimise this is to relate the triggers and desired safe behav-iours to positive identities like pilots, who use checklists dili-gently and professionally.

The last approach to shape the path is to rally the herd or to make safe behaviour contagious. This is focused on the behavioural norms of the target group the change leader is trying to change. Individuals tend to behave in accordance with the norm in a group. This has been supported by research such as Goh and Binte Sa'adon (2015), who found that the norm, as perceived by workers, strongly influences the intention and actual behaviour of a worker hooking on his safety harness.

The elephant always looks to the herd for cues about behaviour. Workers with unsafe behaviours can be identified through inspections and behavioural observations. Once these workers are identified, supervisors and change leaders can get them to change by assigning buddies with safe behavioural habits to the worker to remind the worker to work safely at the beginning of and during shifts. The intention is to work through the smallest possible group to influence the individual's behaviour to bring about positive cultural change. Concurrently, individuals identified for their safe behaviour will need opportunities to share their experience in implementing the desired change and supporting each other in the effort.

Heath and Heath's (2010) Rider-Elephant-Path model is a compilation of different research on change management. The model is simple and practical for change at individual and organisational levels. Not all the nine approaches highlighted are always useful, and overlaps exist between the different approaches, so change leaders need to consider the situation and use the model selectively.

8.4.3 Senge's change model

Another useful literature on change management is (Senge, 1999), which highlighted three basic virtuous cycles that need to be in place for organisational change to be implemented and sustained (Fig. 8.2):

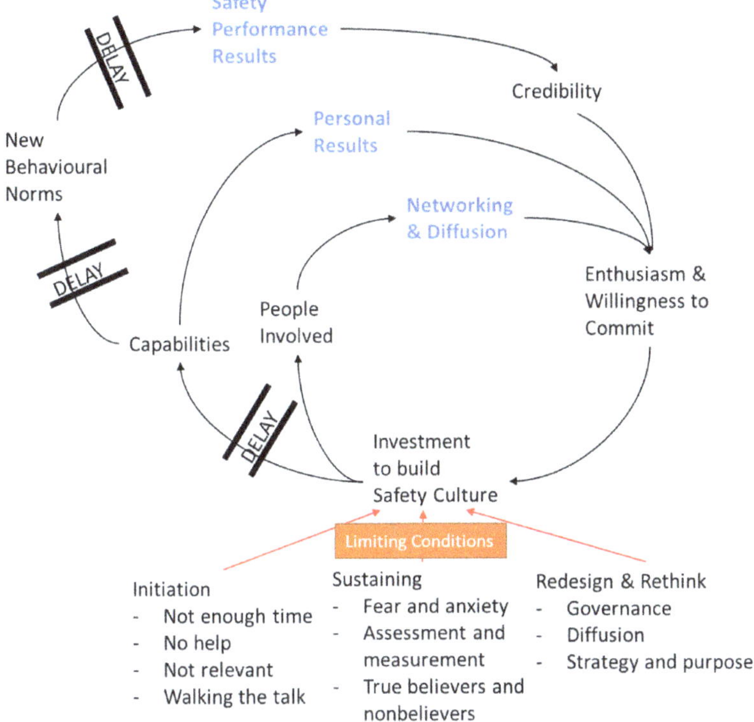

Fig. 8.2 Virtuous cycles and limiting conditions. (Adapted from (Senge, 1999)).

1. **Organisation's Safety Performance Results:** The investment to build safety culture will take some time (delay) before capabilities in WSH management and risk controls can be improved. After a sustained period, the behaviours of people in the organisation will improve, and it will become a new norm for people to work safely. Since unsafe acts and conditions will be reduced, managers should expect to see an improvement in WSH performance. The improvement will improve the credibility of the initial investments and interventions. This will encourage more investment in the programme and spark a virtuous improvement cycle.

2. **Networks of Committed People:** The initial investment to build an organisation's safety culture will involve people in improving WSH. The directly involved people will typically be more committed, but their initial numbers may be small. Getting these early supporters to network and diffuse their commitment and support will help to promote enthusiasm for the programme. This will encourage stronger commitment to the safety culture programmes and further investment of resources and time.

    This virtuous cycle will sustain the programme and keep people engaged.

3. **Personal Results:** The initial investment to build safety culture will increase the organisation's WSH capabilities, and after some time, the organisation should ensure that individuals see personal results from their improved WSH management capabilities. This means that the way that organisations measure and reward performance needs to encourage good safety behaviour. Once individuals see the positive consequences of their involvement in the safety culture programme, they will be more enthusiastic and willing to commit to the programme. This will then develop into a virtuous cycle at the personal level.

All three virtuous cycles will help organisations to reap the benefits of their safety culture improvement programme. Managers trying to get organisations to change their safety culture or implement organisational changes will need the help of these virtuous cycles.

However, for the organisational change to be successful, the organisation must remove barriers to the three virtuous cycles (see Fig. 8.2). The barriers and solutions identified by Senge (1999) are summarised in Table 8.1.

Table 8.1   Summary of the key change barriers and solutions. (Adapted from (Senge, 1999)).

Barriers or challenges	Possible solutions
Initiation	
1. "We don't have time for this stuff" • Lack of control over one's time. People involved in change initiatives need enough flexibility to devote time to reflect and practise • People might be committed, but their lack of time creates an inability to meet the demands of the change initiative • People can end up frustrated and giving up on the initiative • Especially true for pilot team members initiating change	• Integrate initiatives • Proper scheduling • Trust people to use their own time • Value unstructured time • Build productivity capability • No politics or games-playing and nonessential demands
2. "We have no help!" • Inadequate coaching, guidance and support for innovating group • No assistance to help build capacity to sustain change	• Invest in help — consultants, internal experts, guides, etc. • Creating capacity for coaching • Find a partner to confide in • Building coaching into line management • Treat seeking help as norm and positive
3. "This stuff isn't relevant!" • The absence of a clear, compelling business case for learning and change • Resulting in a lack of commitment and enthusiasm	• Build awareness among key leaders • Explicitly raise questions about relevance in the pilot group • Make more information available to pilot group members — connect to individual and job goals

(Continued)

Table 8.1 (*Continued*)

Barriers or challenges	Possible solutions
	• Keep training linked tightly to business results. • Revisit relevance periodically
4. "They're not walking the talk!" • Lack of clarity and credibility of the management's aims and values • Leads to a lack of trust in management to carry out change	• Develop espoused aims and values that are credible • Demonstrate the values and commitment to aims through actions • Work with partners • Cultivate patience under pressure and with bosses • Develop a greater sense of organisational awareness (go to the ground) • Reflect on your beliefs about people and discuss values
Sustaining	
5. Fear and anxiety — Common questions: • Am I safe? • Am I adequate? • Can I trust myself and others to sustain the change?	• Start small and build momentum before confronting difficult issues • Avoid frontal assaults • Set an example of openness • Learn to see diversity as an asset • Use defects as opportunities for learning • Make sure participation in pilot groups and change initiatives is by choice and not by coercion • Remember fear and anxiety are natural; forcing creates more anxiety

Table 8.1 (*Continued*)

Barriers or challenges	Possible solutions
6. Assessment and measurement • Expectations of results • Delay in seeing results • Negative results due to change • Lack of measurement of results	• Appreciate time delays in change initiatives • Assign suitable metrics (not just cost or outcome) • Recognise and appreciate progress
Redesign and rethink	
7. True believers vs. nonbelievers (lack of understanding between groups) • Pilot group (believers) vs. those not in the pilot (nonbelievers) • Fear and anxiety arise among those not involved in the change • Lack of engagement between groups • Arrogance among pilot group members who think they are always right • Pilot group might feel under-appreciated for their efforts	• Become "bicultural" — incubators and mainstream become distinct groups • Mentoring of line leaders • Build the ability to engage to spread change initiatives beyond the pilot group • Cultivate reflective openness • Respect people's inhibitions about personal change • You don't have to convince people • Remove jargon and keep the language simple • Lay a foundation of transcendent values (tolerance and flexibility)
8. Governance ("Who is in charge of this stuff?") • How to institutionalise the change initiatives to be aligned with existing structures • Balance between autonomy for change and corporate control	• Pay attention to boundaries and be strategic when crossing them • Articulate the case for change in terms of business results • Make executive leaders' priorities part of your creative thinking

(Continued)

Table 8.1 (*Continued*)

Barriers or challenges	Possible solutions
	• Experiment with cross-functional, cross-boundary teams if you can get them sponsored by the hierarchy • Begin at the beginning: with governing ideas • Deploy new rules and regulations judiciously • Never underestimate the power of small changes in complex situations — if they are the "right" changes • Be prepared for a long journey, and do not embark alone
9. Diffusion ("Reinventing the wheel") • Not learning from pilot groups' experience • Other people's innovations were dismissed as not being applicable to them • Lack of interest • Impatience in repeating change initiatives • Arrogance — Assume that there is nothing new to learn	• Increase the quality and number of coaches • Increase permeability of intra-organisational boundaries • Improve information infrastructure — share innovations and changes • Build learning culture and communities of practice • Research as part of executive accountability
10. Strategy and purpose • Change can create questions about existing strategy and purpose • What do we want to create? • How will it contribute to others? • New ideas on strategy and purpose may not be accepted	• Use scenario thinking to investigate blind spots, signals of unexpected events and organisational purpose • Engage people in organisational strategy and purpose • Expose and test assumptions behind strategy

8.5 Change management plan

Having a change management plan with detailed documentation is useful to guide and manage the organisational change process. An example can be found in the Queensland Government Chief Information Office (n.d.), which includes the following components:

1. Change identification: Type, reason, scope, current status, and desired state
2. Change specification: Policy, structure, processes, relationships, culture, people, information, cost, and risk assessment
3. Change methodology: Stakeholder analysis (target groups, advocates, early adopters, and resistance and drivers)
4. Implementation strategies: Action plan, schedules, communication plan, training plan, IT/business systems plan, and resistance management plan

Documenting the change management plan for a safety culture change initiative is helpful because it helps managers overseeing the change to stay focused and enables them to make changes along the way. A structured approach to safety culture change management is a critical success factor.

8.6 Case study: Alcoa

This case study is adapted from (Duhigg, 2014, Chap. 4), which describes how Alcoa, formerly known as The Aluminum Company of America, successfully established a culture of excellence by focusing on safety as a "keystone habit". Alcoa is a global industrial leader with businesses in different industries. In 1987, Paul O'Neill was appointed as the new Chief Executive Officer (CEO). He was a former government bureaucrat and was relatively unknown in the Wall Street community. O'Neill made a shocking announcement in his first shareholder meeting. He indicated that he intended to make Alcoa the safest company in America, and he wanted to go for zero injuries. This was unprecedented, and investors in the room were shocked.

O'Neill believes that bringing down the injury rates would mean that individuals in the company have agreed to devote themselves to creating a habit of excellence. In this way, safety indicates how good the company is at changing shared habits. O'Neill wanted to attack one habit and then let the effect of the change ripple through the organisation. Duhigg (2014) defined this as a "keystone habit", a habit that has the power to start a chain reaction to influence how all employees behave and communicate in the organisation. These keystone habits are a focus and leverage for managers to initiate fundamental changes in organisational culture. Instead of framing safety culture as an isolated subset of organisational culture, safety culture becomes a trigger for change in organisational culture.

O'Neill required managers to understand why injuries happen. In investigating incidents, managers had to understand how the manufacturing process went wrong to understand and rectify the process problems. Workers need to be educated about quality control, efficient work processes, how to make processes less error-prone, etc. The entire campaign centred around the idea that correct work is safe and vice versa.

O'Neill's commitment was put to the test when a fatality occurred. An extrusion press (a metalworking machine) had stopped operating; a worker jumped over the yellow safety wall surrounding the press and walked across the pit to remove a piece of aluminium jammed into the hinge of a swinging six-foot arm. When the young man removed the jammed aluminium scrap, the machine jumped back into motion, and the mechanical arm swung and struck the worker on his head, crushing his skull and killing him. The worker was a young employee who joined the company a few weeks ago. The worker was eager to take up the job because his wife was pregnant, and they needed the health care the job offered.

O'Neill was informed about the accident in the middle of the night. Within 14 hours, the plant's executives and Alcoa's top officers were ordered to its headquarters to attend an emergency meeting. They reviewed the evidence and analysed the accident in detail. They identified several causes, including:

- two managers saw the worker jump over the barrier but failed to stop him
- a training programme that failed to emphasise to the man that he would not be blamed for a breakdown
- lack of instructions that he should find a manager before attempting a repair
- absence of sensors to automatically shut down the machine when someone stepped into the pit

The CEO said, "We killed this man… It's my failure of leadership… And it's the failure of all of you in the chain of command." This is a very strong statement, but his subsequent actions were more indicative of his commitment. O'Neill gave his home telephone number to workers on the ground, asking them to call him whenever there were safety issues that their managers did not follow up on. According to Duhigg (2014), workers started to call, giving O'Neill great ideas to improve work.

One example is when a worker suggested improving the work process at an aluminium manufacturing plant that manufactured siding for houses. The plant had engaged consultants to choose paint shades to anticipate customers' desired colours. Still, the problem remained unsolved despite the hefty consultant fees. The worker suggested that all the painting machines be grouped so that paint pigments could be switched out faster, allowing them to be nimbler in responding to changes in customers' colour preferences. This simple suggestion resulted in profits on aluminium siding within a year. Interestingly, the worker had had the idea for a decade but had not told his management. Instead, it was only when O'Neill started focusing on improving safety that the communication channel was opened — and that was when the employee thought that his suggestion might be useful and finally decided to raise it. The focus on improving safety culture created a climate for innovation and willingness to change.

By 2000, when O'Neill retired, the annual net income had increased by five times compared to when O'Neill had initially started. Market capitalisation had risen by US$27 billion, and Alcoa had become one of the safest companies in the world. More specifically, before 1987, the Alcoa plant had at least one accident per week, but after O'Neill took over as CEO, some plants did not have accidents that caused more than one day of work to be lost. The company's worker injury rate was one-twentieth of the US average.

This case study shows that improving safety culture is possible, and Alcoa is a bright spot that should be cloned. The importance of change leaders cannot be overstated, and it is obvious that the change was difficult, but the benefits go beyond WSH.

8.7   Conclusions

Improving safety culture is a challenging task. Managers need to structure the change systematically and deliberately. This chapter highlighted several approaches for improving safety culture. Safety culture intervention is a specific type of organisational change; hence, understanding a range of organisational change management models is useful. When managers and change leaders try to improve safety culture, the organisation must learn the new culture. Therefore, Senge's (2006) model of learning organisation, which involves systems thinking, personal mastery, mental model, shared vision, and team learning and dialogue, is a possible guiding framework for the change initiative. This chapter also introduced other change management models, including Lewin's model, Heath and Heath's (2010) Rider-Elephant-Path model, and Senge's (1999) virtuous cycles and barriers. A template for a change management plan was also briefly discussed. Lastly, Alcoa's successful change initiative was presented as a bright spot of how organisational excellence was improved through a safety improvement campaign.

Review questions

1. What are the three general approaches to improving safety culture?
2. Compare the five disciplines of a learning organisation with Reason's model of safety culture. What are the similarities and differences?
3. Compare Heath and Heath's (2010) model of change with Lewin's model. What are the similarities and differences?
4. Describe the three virtuous cycles needed to make sustainable organisational changes.
5. According to Senge (1999), what are the ten barriers to change?
6. Apply the Rider-Elephant-Path model to Alcoa's case study. What were the approaches used to implement the change?

References

Cummings, S., Bridgman, T., and Brown, K. G. (2015). Unfreezing change as three steps: Rethinking Kurt Lewin's legacy for change management. *Human Relations*, **69**(1), 33–60.

Duhigg, C. (2014). *The power of habit: Why we do what we do in life and business.* New York: Random House.

Goh, Y. M. and Binte Sa'adon, N. F. (2015). Cognitive factors influencing safety behavior at height: A multimethod exploratory study. *Journal of Construction Engineering and Management,* **141**(6).

Heath, C. and Heath, D. (2010). *Switch: How to change things when change is hard.* New York: Broadway Books.

Kotter, J. P. and Cohen, D. S. (2002). *The heart of change: Real-life stories of how people change their organizations.* Boston, MA: Harvard Business School Press.

Queensland Government Chief Information Office. (n.d.). Change management plan workbook and template. http://www.nrm.wa.gov.au/media/10528/change_management_plan_workbook_and_template.pdf (accessed: 12 March 2018).

Senge, P. M. (1999). *The dance of change: The challenges of sustaining momentum in learning organizations: A fifth discipline resource.* London: Nicholas Brealey Pub.

Senge, P. M. (2006). *The fifth discipline: The art and practice of the learning organization* (revised and updated). New York: Currency.

CHAPTER 9

Workplace safety and health legislations

9.1 Introduction

In most countries, workplace safety and health (WSH) management is regulated by a suite of regulations. Thus, as part of the WSH management system, the organisation must have procedures to identify and access relevant WSH legislations and ensure compliance with the regulations (see Clause 6.1.3 of ISO 45001:2018 and Chap. 6). This chapter provides an overview of the Singapore Workplace Safety and Health Act (WSHA) and its subsidiary legislations, which have requirements similar to those of WSH legislations in other countries. This chapter relies heavily on the content on the Ministry of Manpower (MOM) website and Singapore Statutes Online.

Table 9.1 contains the list of legislations under the WSHA (Cap. 354A). The WSHA is discussed in Chap. 1. In addition, some of the subsidiary regulations, such as the Workplace Safety and Health Design for Safety Regulations 2015 (S428/2015) (Chap. 5) and the Workplace Safety and Health (Risk Management) Regulations (Cap. 354A, RG 8) (Chap. 4), are also discussed in the relevant chapters. This chapter will briefly discuss other details of the WSHA and selected regulations.

An important point to note is that even though other regulations could cover specific hazards, e.g., structural safety is also

Table 9.1   List of WSHA-related legislations in Singapore.

#	Title	Number
1	Workplace Safety and Health Act	Cap. 354A
2	Factories (Registration and Other Services — Fees and Forms) Regulations	RG 5
3	Factories (Safety Training Courses) Order	Cap. 354A/104, O 12
4	Factories (Work of Engineering Construction) Order	Cap. 354A/ 104, O 6
5	Workplace Safety and Health (Abrasive Blasting) Regulations 2008	S 607/2008
6	Workplace Safety and Health (Asbestos) Regulations 2014	S 337/2014
7	Workplace Safety and Health (Composition of Offences) Regulations	RG 6/2007
8	Workplace Safety and Health (Confined Spaces) Regulations 2009	S 462/2009
9	Workplace Safety and Health (Construction) Regulations 2007	S 663/2007
10	Workplace Safety and Health (Design for Safety) Regulations 2015	S 428/2015
11	Workplace Safety and Health (Exemption) Order	O 1
12	Workplace Safety and Health (Explosive Powered Tools) Regulations 2009	S 325/2009
13	Workplace Safety and Health (First-Aid) Regulations	RG 4
14	Workplace Safety and Health (General Provisions) Regulations	RG 1
15	Workplace Safety and Health (Incident Reporting) Regulations	RG 3
16	Workplace Safety and Health (Major Hazard Installations) Regulations 2017	S 202/2017

Table 9.1   (*Continued*)

#	Title	Number
17	Workplace Safety and Health (Medical Examinations) Regulations 2011	S 516/2011
18	Workplace Safety and Health (Noise) Regulations 2011	S 424/2011
19	Workplace Safety and Health (Offences and Penalties) (Subsidiary Legislation under Section 66(14)) Regulations	RG 5
20	Workplace Safety and Health (Operation of Cranes) Regulations 2011	S 515/2011
21	Workplace Safety and Health (Registration ofFactories) Regulations 2008	S 501/2008
22	Workplace Safety and Health (Risk Management) Regulations	RG 8
23	Workplace Safety and Health (Safety and Health Management System and Auditing) Regulations 2009	S 607/2009
24	Workplace Safety and Health (Scaffolds) Regulations 2011	S 518/2011
25	Workplace Safety and Health (Shipbuilding and Ship-Repairing) Regulations 2008	S 270/2008
26	Workplace Safety and Health (Transitional Provision) Regulations	RG 7
27	Workplace Safety and Health (Work at Heights) Regulations 2013	S 223/2013
28	Workplace Safety and Health (Workplace Safetyand Health Committees) Regulations 2008	S 355/2008
29	Workplace Safety and Health (Workplace Safety and Health Officers) Regulations	RG 9
30	Workplace Safety and Health (Workplaces Subject to Act) Order 2007	S 72/2007

covered under the Building Control Act, the requirements in the WSHA and its subsidiary regulations still apply even when other legislation is complied with.

9.2 Workplace Safety and Health Act (1st November 2022 version)

The key duties of the different duty holders were discussed in Chap. 1. The following will provide additional information highlighted in the WSHA.

9.2.1 Difference between workplace and factory

The WSHA provided a distinction between a "workplace" and a "factory". A "workplace" is any place where a person is working, will be working for a short time, or usually works. Some of these workplaces are further classified as factories. A factory is a term carried over from the repealed Factories Act, which was the main legislation regulating industrial safety and health before the WSHA. A factory is any premises in which any of the following is carried out:

(a) The making of any article or part of any article.
(b) The alteration, repair, ornamentation, finishing, cleaning, or washing of any article.
(c) Breaking up or demolishing any article.
(d) Adapting any article for sale.

As described in the Fourth Schedule of the WSHA, examples of factories include manufacturing plants, car-servicing workshops, shipyards, and construction worksites. These workplaces are considered higher risk and are subjected to additional regulations, e.g., periodic audits and a mandatory WSH committee.

9.2.2 Protecting employees

The WSHA Section 18 protects employees by stating that employers shall not "(1)(a) deduct, or allow to be deducted, from the sum contracted to be paid by him to any employee of his; or (b) receive, or allow any agent of his to receive, any payment from any employee of his, in respect of anything to be done or provided by him in accordance with this Act in order to ensure the safety, health or welfare of any of his employees at work". Anecdotally, the author has heard of practices in the industry where employers (including principals) deduct the pay of employees or ask for payment for personal protective equipment (PPE) that should have been provided to the workers as required by WSHA. In addition, Section 18 also makes it illegal for employers to dismiss or threaten to dismiss an employee if the employee whistle-blows on the employer for WSHA-related contraventions or for performing his duties as a member of the WSH committee.

9.2.3 Powers of the Commissioner

The WSHA gives the Commissioner for WSH, who is typically the Director of the Division of Occupational Safety and Health in MOM, powers to issue stop-work orders (to stop the work at a workplace) and remedial orders (to improve the conditions in a workplace). Failure to comply with a remedial order can result in a fine not exceeding S$50,000 or imprisonment for a term not exceeding 12 months or both. In the case of a continuing offence, the duty holder may have to pay a further fine not exceeding S$5,000 for every day or part thereof during which the offence continues after conviction. On the other hand, failure to comply with a stop work order can result in a fine not exceeding S$500,000 or imprisonment for a term not exceeding 12 months or both. In the case of a continuing offence, the duty holder may have to pay a further fine not exceeding S$20,000 for every day or part thereof during which the offence continues after conviction.

The Commissioner also has the power to enter a workplace for the purpose of inspection and investigation. The WSHA also made provisions to protect accident-related evidence, where any alteration or addition to machinery, equipment, etc., related to an accident and dangerous occurrence is an offence.

Section 27A of the WSHA is a proactive amendment of the WSHA in 2018. The section allows the Commissioner to prepare and publish a learning report on any accident, dangerous occurrence, or occupational disease in a workplace that is still being investigated. Prior to this section, it was difficult for MOM to disseminate important WSH information gained from investigations prior to prosecution because the information released may influence the outcome of the prosecution. The learning report may:

(a) contain an account of the accident, dangerous occurrence or occupational disease;

(b) specify the cause or causes of, and circumstances or factors leading to, the accident, dangerous occurrence or occupational disease insofar as they may be ascertained;

(c) contain an opinion by a person with technical or specialised knowledge of the machinery, equipment, plant, article, process, substance, work or workplace involved in the accident, dangerous occurrence, or occupational disease;

(d) contain a warning of any danger or risk to the safety and health of persons at work or persons who may be affected by any undertaking carried on in the workplace;

(e) contain any recommendation to prevent or minimise the recurrence of any similar accident, dangerous occurrence or occupational disease in a workplace;

(f) contain any other matter that the Commissioner considers relevant, taking into account the sole objective mentioned in subsection (2), which is "to prevent or minimise the recurrence of any accident, dangerous occurrence or occupational disease in a workplace, and not to apportion blame or liability".

The learning report is not admissible in any legal proceedings. This is aligned with the approach used by the US Chemical Safety Board (CSB), where the investigation report written by CSB is not admissible in court, and the US Occupational Safety and Health Administration (OSHA) conducts an independent investigation for its prosecution purposes. However, as of 10 November 2023, there are only two learning reports on the MOM's website. More needs to be done to increase the number of learning reports.

9.2.4 Safety and health management arrangements

Part VII of the WSHA specifies requirements for safety and health management arrangements. These requirements pertain to WSH Officers (WSHOs) and coordinators, WSH committees, WSH auditors, safety and health training courses, and the need for the Commissioner's approval for people to perform statutory roles (authorised examiner for dangerous machines, WSHO, WSH coordinator, WSH auditor, and accredited training provider). The WSHA only provides the framework to regulate each of these areas of safety and health management arrangements. The details of the actual requirements are usually found in relevant subsidiary regulations and government gazettes.

9.2.5 Workplace Safety and Health Council and approved code of practice

The WSHA also provided for the establishment of the WSH Council. According to Section 40A, the functions of the WSH Council are:

(a) to develop or facilitate the development of acceptable practices relating to safety, health, and welfare at work;

(b) to promote the adoption of acceptable practices relating to safety, health, and welfare at work;

(c) to devise, organise and implement programmes and other activities for or related to providing support, assistance or advice to any person or organisation in preserving, improving and promoting safety, health, and welfare at work;

(d) to facilitate and promote the development and upgrading of competencies, skills, and expertise of the workforce relating to safety, health, and welfare at work;

(e) to research into any matter relating to safety, health, and welfare at work;

(f) to grant prizes and scholarships and to establish and subsidise lectureships in universities and other educational institutions in subjects relating to safety, health, and welfare at work;

(g) to provide practical guidance with respect to the requirements of this Act relating to safety, health, and welfare at work;

(h) to do all the things that it is authorised or required to do under the WSHA.

Sections 40B and 40C then stipulated the role of the codes of practice as issued or approved by the WSH Council. The codes of practice approved by the WSH Council are frequently Singapore Standards published by the Standards, Productivity and Innovation Board under Section 7(2)(h) of the Standards, Productivity and Innovation Board Act (Cap. 303A). Essentially, the approved code of practice (ACoP) is a non-mandatory benchmark for the court to determine whether a measure is reasonable and practicable. In the absence of a suitable ACoP, relevant Singapore Standards not approved by the WSH Council and industry guidelines can also be used to establish reasonable and practicable standards.

9.2.6 Schedules

The First Schedule provided the types of dangerous occurrences that must be reported to the MOM. They include bursting failures of machinery, crane collapses, significant explosions or fires, significant electrical short-circuits or failures, explosions or failures of a steam boiler's structure(s), failures or collapses of formwork or its supports, scaffold (15 metres in height) collapses of suspended or hanging scaffolding above two metres, and accidental seepage or entry of seawater into a dry dock.

The Second Schedule identified 35 occupational diseases that must be reported to the MOM:

1. Aniline poisoning	13. Diseases caused by excessive heat	25. Organophosphate poisoning
2. Anthrax	14. Hydrogen sulphide poisoning	26. Phosphorus poisoning
3. Arsenical poisoning	15. Lead poisoning	27. Poisoning by benzene or a homologue of benzene
4. Asbestosis	16. Leptospirosis	28. Poisoning by carbon monoxide gas
5. Barotrauma	17. Liver angiosarcoma	29. Poisoning by carbon disulphide
6. Beryllium poisoning	18. Manganese poisoning	30. Poisoning by oxides of nitrogen
7. Byssinosis	19. Mercurial poisoning	31. Poisoning from halogen derivatives of hydrocarbon compounds
8. Cadmium poisoning	20. Mesothelioma	32. Musculoskeletal disorders of the upper limb
9. Carbamate poisoning	21. Noise-induced deafness	33. Silicosis
10. Compressed air illness or its sequelae, including dysbaric osteonecrosis	22. Occupational asthma	34. Toxic anaemia
11. Cyanide poisoning	23. Occupational skin cancers	35. Toxic hepatitis
12. Diseases caused by ionising	24. Occupational skin diseases	

The Third Schedule provided a list of activities classified as Work of Engineering Construction, i.e. construction work, while the Fourth Schedule identified 19 categories of workplaces defined as factories. The Fifth Schedule listed 11 types of machinery and equipment (e.g., scaffolds, cranes, forklifts, bar benders, and welding machines) and 17 types of hazardous substances (e.g., corrosive, flammable, explosive, oxidising, toxic, and carcinogenic substances) that require their manufacturers and suppliers to ensure compliance with Section 16 of WSHA, which make it compulsory to:

(a) provide information on health hazards and how to safely use the machinery, equipment, or hazardous substance;
(b) examine and test the machinery, equipment, or hazardous substance to ensure that it is safe for use;
(c) provide results of any examinations or tests of the machinery, equipment, or hazardous substances.

The persons who erect, install or modify the machinery and equipment identified in the Fifth Schedule are required under Section 17 to ensure that the method of work is in accordance with the manufacturer's instruction and that the machinery or equipment is safe for use after installation, erection or modification.

9.3 Workplace Safety and Health (Asbestos) Regulations 2014

Asbestos is a hazardous substance that can cause lung diseases such as asbestosis (scarring and fibrosis of the lung tissues), mesothelioma (a cancer of the chest and abdominal lining), and lung cancer (Workplace Safety and Health Council, 2017). The substance is not a specific mineral but a group of fibrous silicate minerals (Tranter, 2004) commonly used for insulation, fire protection, and acoustic purposes. The Workplace Safety and Health (Asbestos) Regulations 2014 define asbestos as crocidolite, actinolite, anthophyllite, amosite, tremolite, chrysotile, amphiboles, or a mixture containing any such minerals. The material has been banned in Singapore since 1989, but it can still be found in the thermal insulation of pipes in vessels, plants, furnaces, and buildings constructed before 1989. The National Environment Agency (NEA) and the MOM regulate the use and disposal of asbestos in Singapore. The MOM focuses on the protection of people at work, and the key legal instrument is the Workplace Safety and Health (Asbestos) Regulations 2014 ("Asbestos Reg").

The Asbestos Reg contains six parts and a Schedule. The Schedule defines the material, substance, product, or article classified as "specified material" that requires special attention and measures under the Asbestos Reg. The Schedule listed the following "specified material":

1. Cable penetration insulation
2. Fire protection board, panel, wall and door
3. Gasket
4. Refractory lining
5. Sprayed insulation
6. Thermal insulation of pipe, boiler, pressure vessel, and process vessel

For any work that involves the above-specified material likely to contain asbestos or any demolition, alteration, addition or repair of a building built before 1 January 1991, the employer or principal must engage a competent person to assess if the specified material contains asbestos. The competent person will then survey to collect suitable samples and send them for testing by an approved testing body. A survey report must then be provided after the test concerning the presence of asbestos.

If asbestos is present in the specified material, then the Occupier must ensure that the asbestos removal work is conducted by an Approved Asbestos Removal Contractor (AARC) (a list of AARCs can be found on the MOM website). Work involving asbestos must be conducted by workers trained to understand the hazards of asbestos and the safe way to handle the material, including training on the use of respiratory protective equipment. These workers will also be subjected to medical examinations stipulated in the WSH (Medical Examinations) Regulations 2011. The AARC must engage a competent person to develop an asbestos-removal plan of work and supervise the asbestos-removal work. The AARC must also ensure the implementation of the plan. The MOM must be notified of all asbestos removal work seven days before the commencement of the removal work.

9.4 Workplace Safety and Health (Confined Spaces) Regulations 2009

According to the Workplace Safety and Health (Confined Spaces) Regulations 2009 ("Confined Spaces Reg"), a confined space is defined as any chamber, tank, man-hole, vat, silo, pit, pipe, flue, or other enclosed space in which:

(a) dangerous gases, vapours or fumes are liable to be present to such an extent as to involve a risk of fire or explosion or persons being overcome thereby;
(b) the supply of air is inadequate or is likely to be reduced to be inadequate for sustaining life or
(c) there is a risk of engulfment by material.

Confined spaces are dangerous because workers may not easily identify them. There have been accidents where multiple workers were killed because when the first worker became unconscious in the confined space, co-workers entering the confined space to rescue the affected worker were also affected by the harmful gases or substances in the space. The Confined Spaces Reg stipulates the requirement for the Occupier to record the description and location of all fixed and stationary confined spaces and inform the relevant people of the hazards of these identified confined spaces. The Occupier must also ensure that the access and egress, entrance, lighting, and ventilation of the confined space(s) are safe and suitable for the confined space(s).

The Confined Spaces Reg stipulates a permit-to-work system where a competent authorised manager and a competent confined space safety assessor are appointed. The confined space entry permit must specify the following:

(a) the description and location of the confined space;
(b) the purpose of entry into the confined space;
(c) the results of the gas testing of the atmosphere of the confined space;
(d) its period of validity.

The Occupier is to evaluate if the confined space entry is indeed necessary. If the entry is necessary, the Occupier must ensure that the permit-to-work system is implemented. If the person entering the confined space is wearing a suitable breathing apparatus, authorised by the authorised manager to enter the confined space, and "where reasonably practicable, is wearing a safety harness with a rope securely attached and there is a confined space attendant keeping watch outside the confined space which is provided with the means to pull such a person out of the confined space in an emergency", then the confined space entry permit is not necessary.

The application for the confined space entry permit is to be made by the supervisor of the person entering the confined space. The submitted application form will have to contain the safety measures to protect the worker(s) entering the confined space. These safety measures will have to be assessed by the confined space assessor and then approved by the authorised manager before the workers can enter the confined space. The entry permit must be displayed at the entrance of the confined space.

One of the duties of the confined space entry assessor is to test the atmosphere of the confined space for oxygen content (19.5% to 23.5% by volume), level of flammable gases or vapour (less than 10% of its lower explosive limit), and the presence of toxic gas or vapour (that it does not exceed the permissible exposure levels as specified in the First Schedule to the WSH (General Provisions) Regulations).

The gas meter used in the testing must be properly calibrated. The confined space safety assessor will have to conduct periodic testing of the atmosphere during the conduct of the work. In addition, if there is more than one person present in the confined space, at least one of them must have a suitable gas detector that continuously monitors the atmosphere in the confined space. In the event that a hazardous atmosphere is detected, all personnel must vacate the confined space. A re-evaluation of the confined space must be conducted, and a new permit will have to be issued before any worker can enter the confined space.

Another key hazard of a confined space is "incompatible work", which means work that is carried out at or in the vicinity of any work carried out in the confined space and which is likely to pose a risk to the safety and health of persons present in the confined space. A set of incompatible work includes hot work (e.g., welding, hot-cutting and grinding) and spray-painting in confined spaces. The work may be in adjacent confined spaces, but when the incompatible substances come into contact, a major fire and explosion can occur. Thus, the Confined

Spaces Reg requires anyone who is aware of incompatible work to immediately report it to their supervisor, WSH personnel, and authorised manager.

9.5 Workplace Safety and Health (Construction) Regulations 2007

The Workplace Safety and Health (Construction) Regulations 2007 ("Con Reg") is one of the foundational WSH regulations for the construction industry. It contains 16 parts and 142 Regulations.

In addition to the WSH (Safety and Health Management System and Auditing) Regulations 2009, the Con Reg requires the Occupier to convene site coordination meetings to coordinate site activities to ensure the safety, health and welfare of persons at work in the worksite. The site coordination meeting must be chaired by the "project manager" (see Chap. 7), which means the person who is stationed at a worksite and who has overall control of all the works carried out in the work site, and includes any competent person appointed by the occupier of the worksite if the project manager is unable to perform his duties under the Con Reg. In addition, the Occupier of a project with a contract sum of less than $10 million will have to appoint a WSH co-ordinator to assist the occupier in identifying unsafe conditions or unsafe work practices, recommend to the occupier reasonably practicable measures to remedy the unsafe conditions or unsafe work practices, and assist the occupier in implementing the recommended reasonably practicable measures. This role is similar to that of a WSHO, but a WSH coordinator's training level is lower than that of a WSHO. The Con Reg also requires suitable safety and health training for all workers and supervisors.

Similar to the Confined Spaces Reg, the Con Reg also requires a permit-to-work system for the following high-risk construction activities:

(a) Demolition work
(b) Excavation and trenching work in a tunnel or hole in the ground exceeding a depth of 1.5 metres
(c) Lifting operations involving a tower, mobile or crawler crane
(d) Piling work
(e) Tunnelling work

The remaining portions of the Con Reg cover different hazards, such as structural safety and stability, storage and stacking of materials and equipment, falling objects, slipping hazards, protruding hazards, personal protective equipment, electrical safety, cantilevered platforms, formwork, demolition, excavation and tunnelling, piling, cranes, and employee's lifts.

9.6 Workplace Safety and Health (General Provisions) Regulations

The Workplace Safety and Health (General Provisions) Regulations ("GP Reg") covers all workplaces. It covers general provisions relating to health, safety and welfare. The hazards identified under health issues include infectious agents and biohazardous material, overcrowding, ventilation, lighting, drainage (to prevent wet floors), sanitary conveniences, vibration, and excessive heat or cold and harmful radiations.

The range of safety hazards covered in the GP Reg is much wider. Several of the Regulations cover machinery hazards. For example, Regulation 11 stipulates the requirements for occupiers to ensure secure fencing (or guarding) for prime movers (e.g., electric motor and internal combustion engines) and connecting flywheels and moving parts. Regulation 12 further indicates that fencing is not necessary only if the dangerous parts of the machinery are made safe (i.e., no person can be harmed) because of its position, construction, or other means. Another requirement is the need for an emergency cut-off that can efficiently cut off the power to the machinery. Regulation 13 requires safety measures to be implemented to assure the safety of maintenance workers who may need to remove the machinery fencing while the machinery is operating. The worker must be at least 18 years old. He must have been trained for maintenance or repair jobs and be wearing suitable clothing with no loose ends.

Other safety requirements include electrical installation and equipment (prevention of electrical shock) and requirements to ensure fencing and other safeguards are properly constructed, used and maintained. The occupier also needs to set up a set of lock-out procedures to prevent accidental activation or energising during the inspection, cleaning, repair or maintenance of any plant, machinery, equipment, or electrical installation in the workplace. The Regulations also cover the danger of a person falling into the tank, structure, sump or pit containing scalding, burning, corrosive, or toxic liquid. Self-acting machines (e.g., robotic arms) are required to be made safe by, for example, ensuring no person can be exposed to the machine in operation and having warning signs.

Regulations 19 to 21 (hoists and lifts, lifting gears, lifting appliances, and lifting machines) are related to the role of an Authorised Examiner (AE), who is essentially a special class of Professional Engineers authorised by the MOM to conduct the inspection and certification of dangerous lifting-related machinery. The occupiers and AEs play important roles in preventing accidents like lifted objects falling from height and failure of hoists.

Even though there is a Workplace Safety and Health (Work at Heights) Regulations 2013 ("WAH Regs"), the GP Regs contains several requirements on fall from height hazards. Regulation 23 imposes a duty on the occupier to ensure that all openings in the floors of the workplace are securely covered or fenced unless the fencing is impracticable. Regulation 23 requires employers to provide "secure foothold(s) and handhold(s)", as far as is reasonably practicable, for any worker working at 2 metres or higher or if the worker can fall into any substance that can cause drowning or asphyxiation. If secure footholds and handholds are not reasonably practicable, then the employer must provide other suitable means, such as safety harnesses or safety belts (i.e., fall arrest or travel restraint systems). It was stipulated that the anchorage for the harnesses or safety belts must not be lower than the level of the worker's working position.

Other safety hazards covered included storage of goods, explosives, flammable dust, gas, vapours or substances, pressure vessels (e.g., steam boiler, steam receiver, air receiver and refrigerating plant, which require AE inspection and certification), and fire.

Part IV covers toxic dust, fumes or other contaminants, permissible exposure levels of toxic substances, hazardous substances, warning labels, and safety data sheets. It is noted that the occupier must ensure that any hazardous substance found in a workplace must have a safety data sheet. The First Schedule of the GP Reg shows the permissible exposure levels (PELs) for a wide range of hazardous hazards.

9.7　Workplace Safety and Health (Incident Reporting) Regulations

Under the Workplace Safety and Health (Incident Reporting) Regulations ("IR Reg"), the employer is responsible for notifying the Commissioner of Workplace Safety and Health (in practice, this will be through the MOM) of any fatal accidents involving its employees as soon as is reasonably practicable. The occupier will have to report the death of non-workers and the self-employed to the Commissioner. The report must be made within ten days of the accident. If a dangerous occurrence occurs (see First Schedule of the WSHA), the occupier shall notify the Commissioner as soon as is reasonably practicable. This must be done not later than ten days after the occurrence.

As of 31 August 2020, in the event of an injury, the employer must submit a report to the Commissioner if the employee "is certified by a registered medical practitioner or registered dentist to be unfit for work, or to require hospitalisation or to be placed on light duties, on account of the accident" (see Chap. 1), within ten days after the date the employer first has notice of the accident. If the injury was sustained by a person who was not working or self-employed, the occupier must notify the Commissioner.

Regarding occupational diseases listed in the Second Schedule in the WSHA, the employer and registered medical practitioner shall each submit a report to the Commissioner within ten days of the diagnosis.

The notification reports are legal documents, and employers and occupiers must maintain a copy of the reports for at least three years.

9.8 Workplace Safety and Health (Operation of Cranes) Regulations 2011

Cranes are dangerous machines that can kill many people during an accident. Thus, the Workplace Safety and Health (Operation of Cranes) Regulations 2011 ("OOC Reg") identified a list of duties and requirements for the responsible person, who is frequently the occupier. One of the key administrative controls is the lifting plan. The lifting plan is described in more detail in the Code of Practice on Safe Lifting Operations in the Workplaces (Workplace Safety and Health Council, 2014). Cranes exceeding 5 tonnes and tower cranes can only be operated by a registered crane operator. Cranes not exceeding 5 tonnes (e.g., a mini crane or lorry crane) can only be operated by a competent operator, but the operator need not be registered with the Commissioner for WSH.

A registered crane operator is required to:

(a) conduct a pre-start inspection on the crane;
(b) ascertain the ground conditions and report to the lifting supervisor if he feels that it is not safe;
(c) ensure that any outrigger is fully extended and secured when it is required;
(d) carry out lifting operations only if he has been briefed by the lifting supervisor on the lifting plan;
(e) ascertain the weight of the load, and not lift when a signalman is required and a clear signal has not been given;
(f) operate the crane in a safe manner (e.g., no load over public area, no operation within 3 metres of a live overhead power line, no dragging or pulling of a load, and securely block a crane parked on a slope) and report any failure or malfunction of the crane to the lifting supervisor and make a record of the failure or malfunction in the crane's log book or log sheet.

The OOC Regs also identified other lifting personnel, including the lifting supervisor, rigger and signalman. The lifting supervisor is to coordinate all lifting activities, ensure only suitable and appointed personnel are involved in the lifting operations, ensure that the ground conditions are safe, brief all lifting personnel on the lifting plan, and take measures to rectify all unsatisfactory or unsafe conditions to ensure that the lifting operation can be conducted safely.

Only approved crane contractors (ACCs) are allowed to install, repair, alter, or dismantle mobile cranes or tower cranes. ACCs must be competent in their duties, and only companies approved by the Commissioner for WSH can become ACCs. The

ACCs need to be familiar with the crane manufacturer's instructions and manuals, and they must follow the instructions and manuals strictly. The owner of the cranes must ensure that every crane is tested and certified by an AE.

9.9 Workplace Safety and Health (Safety and Health Management System and Auditing) Regulations 2009

The Workplace Safety and Health (Safety and Health Management System and Auditing) Regulations 2009 ("SHMS Reg") provide further details on the requirements for WSH Management Systems (WSHMS) and auditing on WSHMS. In addition, Part II of the SHSM Reg provides details on the registration of WSH auditors.

The Second, Third and Fourth Schedules of the SHMS Reg are summarised in Table 9.2. These high-risk factories are required to put formal WSHMS in place and have periodic audits. In addition, construction sites with a contract sum of S$30 million or more and shipyards with 200 or more employees must conduct an internal review of their WSHMS every six months and 12 months, respectively. The frequencies of audits and internal reviews are generally based on the past WSH performance and risks posed by the different industries. The occupier is required to engage the auditor to audit the WSHMS.

The SHMS Reg also spells out the role of a WSHMS auditor, which includes an audit of any risk assessment (RA) relating to the workplace or the work carried out in that workplace, as well as any work process at the workplace and the workplace. In an audit, the auditor's duties include:

(a) auditing the workplace in such manner as the Commissioner may determine;

Table 9.2   Summary of audit requirements for different types of workplaces in Singapore.

Type of workplace	Definition	Condition to require WSHMS	Condition to require audit by approved auditor	Frequency of audit by approved auditor	Frequency of internal review if an audit is required
1. Construction worksite	Any premises where any building operation or works of engineering construction is or are being carried out by way of trade or for purposes of gain, whether or not by or on behalf of the Government or a statutory body, and includes any line or siding (not forming part of a railway), which is used in connection with the building operation or works of engineering construction.	All worksites	Contract sum of $30 million or more	At least once every six months	At least once every six months
2. Shipyard	Any yard (including any dock, wharf, jetty, quay, and the precincts thereof) where the construction, reconstruction, repair, refitting, finishing, or breaking up of ships is carried out, and includes the waters	All shipyards	200 or more persons are employed	At least once every 12 months	At least once every 12 months

Type of workplace	Definition	Condition to require WSHMS	Condition to require audit by approved auditor	Frequency of audit by approved auditor	Frequency of internal review if an audit is required
	adjacent to any such yard where the construction, reconstruction, repair, refitting, finishing, or breaking up of ships is carried out by or on behalf of the occupier of that yard.				
3. Metalworking factory	Any factory engaged in the manufacturing of fabricated metal products, machinery or equipment.	100 or more persons are employed	100 or more persons are employed	At least once every 12 months	Not applicable
4. Oil and gas plant	Any factory engaged in the processing or manufacturing of petroleum, petroleum products, petrochemicals, or petrochemical products.	All oil and gas plants	All oil and gas plants	At least once every 12 months	Not applicable
5. Storage of hazardous substances	Any premises where the bulk storage of toxic or flammable liquid is carried on by way of trade or for the purpose of gain.	All storage with a capacity of 5,000 or more cubic metres for such toxic or flammable liquid	All storage with a capacity of 5,000 or more cubic metres for such toxic or flammable liquid	At least once every 24 months	Not applicable

(Continued)

Table 9.2　(*Continued*)

Type of workplace	Definition	Condition to require WSHMS	Condition to require audit by approved auditor	Frequency of audit by approved auditor	Frequency of internal review if an audit is required
6. Chemical plant	Any factory engaged in the manufacturing of chemicals.	Any factory engaged in the manufacturing of: (a) fluorine, chlorine, hydrogen fluoride, or carbon monoxide; or (b) synthetic polymers	Any factory engaged in the manufacturing of: (a) fluorine, chlorine, hydrogen fluoride, or carbon monoxide, or (b) synthetic polymers	At least once every 24 months	Not applicable
7. Pharmaceutical plant	Any factory engaged in the manufacturing of pharmaceutical products or their intermediates.	All pharmaceutical plants	All pharmaceutical plants	At least once every 24 months	Not applicable
8. Semiconductor plant	Any factory engaged in the manufacturing of semiconductor wafers.	All semiconductor plants	All semiconductor plants	At least once every 24 months	Not applicable

(b) submitting an audit report to the occupier of the work-
     place upon completion of the audit indicating the findings
     and recommendations in the report;
(c) advising the occupier of the workplace to take immedi-
     ate action to remedy any unsafe condition or unsafe work
     practice found during the audit that may result in immi-
     nent danger to the safety and health of persons at work;
(d) reporting to the Commissioner the unsafe condition or
     unsafe work practice referred to in sub-paragraph if the
     occupier:
     (i) refuses to take immediate remedial action, or
     (ii) needs more than a day to remedy the unsafe condition
          or unsafe work practice.

   The WSHMS auditor and occupier must ensure that they do
not have any conflict of interest to ensure the credibility and
independence of the audit.

9.10 Workplace Safety and Health (Work at Heights) Regulations 2013

Falling from height is one of the most common reasons for
workplace fatalities. Thus, the Workplace Safety and Health
(Work at Heights) Regulations 2013 ("WAH Reg") is an important legislation that all workplaces must consider. The WAH Reg
follows the hierarchy of control, which advocates the importance of avoidance of WAH whenever possible. The occupier of
the following workplaces must have a fall prevention plan:

(a) any worksite;
(b) any shipyard;
(c) any factory engaged in the processing or manufacturing of
     petroleum, petroleum products, petrochemicals, or petro-
     chemical products;
(d) any premises where the bulk storage of toxic or flamma-
     ble liquids is carried on by way of trade or for the purpose

of gain and which has a storage capacity of 5,000 or more cubic metres for such toxic or flammable liquids;

(e) any factory engaged in the manufacturing of:

a. fluorine, chlorine, hydrogen fluoride, or carbon monoxide, or

b. synthetic polymers;

(f) any factory engaged in the manufacturing of pharmaceutical products or their intermediates;

(g) any factory engaged in the manufacturing of semiconductor wafers;

(h) any factory not falling within any of the classes of workplaces described in paragraphs 1 to 7 (above) and in which 50 or more persons are employed.

The fall prevention plan is described in the Code of Practice for Working Safety at Heights (Workplace Safety and Health Council, 2013).

The responsible person (employer and/or principal) of any person conducting WAH must ensure that the person is trained in WAH. At the same time, the WAH must be under the "immediate supervision" of a competent person. This means, to this author, that the WAH supervisor should be able to intervene immediately if the workers are working unsafely.

The occupier must ensure that all open sides or openings that can cause persons to fall 2 metres or more are covered or guarded by effective guard rails or barriers to prevent falls. If the guard rails or barriers are removed to facilitate work, they must be reinstated as soon as possible, and the workers exposed to falls from height due to the removal of the guard rails or barriers must be protected by a travel restraint (worker is restrained from reaching the edge or opening) or a fall arrest system (worker's fall can be arrested before he hits the ground or obstacle). The occupier must also ensure that the top guard rail is at least one metre tall and the vertical distance between any two adjacent guard rails is less than 600 mm wide. The guard rail or barrier must also be of good construction, sound material, and adequate strength to withstand the impact during the course of the fall.

The design of fall arrest and travel restraint systems can be based on Singapore Standard 607:2015 Specification for the Design of Active Fall Protection Systems (SPRING Singapore, 2015). The fall arrest system must be safe for use, i.e., the force applied on the user's body must be less than 6 kN, and the user must not hit any obstacle or the ground during the fall. The responsible person overseeing the WAH must appoint a competent person to inspect the anchorage and anchorage line of the travel restraint and fall arrest systems. The inspection by the competent person must be conducted at the start of every shift.

The occupier must ensure that there is always a safe means of access and egress. This includes the staircases that must be effectively barricaded to prevent falls of more than two metres. The openings between the platform and a hoist (aka teagle) must also be securely fenced and provided with secure handholds at the opening or doorway.

Working on roofs, fragile surfaces, and ladders is a highly dangerous form of WAH, and they are highlighted in the WAH Reg. For a fixed vertical ladder above nine metres, there must be an intermediate landing place to ensure that the continuous vertical distance does not exceed nine metres.

"Hazardous work at height" refers to work:

(a) in or on an elevated workplace from which a person could fall a distance of more than 3 metres;
(b) in the vicinity of an opening through which a person could fall a distance of more than 3 metres;

(c) in the vicinity of an edge over which a person could fall a distance of more than 3 metres;
(d) on a surface through which a person could fall a distance of more than 3 metres;
(e) in any other place (whether above or below ground) from which a person could fall a distance of more than 3 metres.

All hazardous WAH requires a permit-to-work system to be set up and implemented by the occupier.

The WAH Reg also regulates industrial rope access systems, making requirements like having two independent anchorage lines compulsory and requiring those responsible to engage a professional engineer to design the anchorage and anchorage line of an industrial rope access system.

9.11 Workplace Safety and Health (Workplace Safety and Health Committees) Regulations 2008

The Workplace Safety and Health (WSH Committees) Regulations 2008 ("WSH Committees Regs") stipulates the formation, meetings and functions of a WSH committee, as highlighted in Section 29(1) of the WSHA. The WSH Committee is required in all factories (high-risk workplaces, as defined in the WSHA), and the occupier of the factory must set it up. The WSH committee must be chaired by a competent person appointed by the occupier. The secretary is usually the WSHO, and there must be members who represent employees and the management. There must always be more employee representatives in the WSH committee than employer representatives.

The WSH committee must meet at least once a month during office hours to discuss WSH matters, and the meetings must be recorded in minutes. The committee must conduct inspections at least once a month. The inspection findings must then be discussed and recorded. Follow-up actions must be implemented. The WSH committee must also conduct an inspection after any accident or dangerous occurrence, and the WSHO must investigate the incident and furnish an investigation report to the WSH committee chairman. The post-incident inspection of the WSH committee must be discussed at a meeting, and a report recording the WSH issues and recommendations must be produced. The occupier must then evaluate the report and take the necessary actions.

The WSH committee also has other functions, such as organising promotional activities to improve WSH and setting WSH guidelines for the factory. For the WSH committee to be effective, the members must be trained in WSH. The WSH Committee has the following powers to:

(a) enter, inspect and examine the workplace at any reasonable time;
(b) inspect and examine any machinery, equipment, plant, installation, or article in the workplace;
(c) require the production of workplace records, certificates, notices, and documents kept or required to be kept under the Act, including any other relevant document, and to inspect and examine any of them;
(d) make such examination and inquiry of the workplace and of any person at work at that workplace as may be necessary to execute his duties;
(e) assess the levels of noise, illumination, heat, or harmful or hazardous substances in the workplace and the exposure levels of persons at work therein;
(f) investigate any accident, dangerous occurrence or occupational disease that occurred within the workplace.

9.12 Workplace Safety and Health (Workplace Safety and Health Officers) Regulations

A WSHO must have completed the WSQ Specialist Diploma in WSH and have at least two years of practical experience related to WSH. In addition, the WSHO must also pass an interview conducted by the Commissioner (MOM). The following work-places must employ a WSHO:

(a) shipyards in which any ship, tanker and other vessels are constructed, reconstructed, repaired, refitted, finished, or broken up;
(b) factories used for processing petroleum or petroleum products;
(c) factories in which building operations or works of engineering construction of a contract sum of S$10 million or more are carried out;
(d) any other factories in which 100 or more persons are employed, except those which are used for manufacturing garments.

The duties of a WSHO include:

(a) assisting the occupier of the workplace or other person in charge of the workplace to identify and assess any seeable risk arising from the workplace or work processes therein;
(b) recommending to the occupier of the workplace or other person in charge of the workplace reasonably practicable measures to eliminate any foreseeable risk to any person who is at work in that workplace or may be affected by the occupier's undertaking in the workplace;
(c) where it is not reasonably practicable to eliminate the risk referred to in subparagraph (b), recommending to the occupier of the workplace or other person in charge of the workplace:
    a. such reasonably practicable measures to minimise the risk, and
    b. such safe work procedures to control the risk;
(d) assisting the occupier of the workplace or other person in charge of the workplace to implement the measure or safe work procedure referred to in subparagraph (b) or (c), as the case may be.

As can be observed, the WSHO is predominantly meant to assist and advise the occupier. To perform their duties, the WSHO is given the same powers as the WSH committee.

9.13 Conclusions

This chapter provided further details of the WSH Act and selected subsidiary regulations. As can be observed, there is a comprehensive WSH framework in Singapore. This WSH framework is not unique to Singapore; many countries have similar WSH frameworks and requirements. A comprehensive legal framework is a critical success factor for any country trying to improve its WSH, but it is still the actual implementation and enforcement of the framework that will have an actual impact on industries, organisations, managers, and ultimately workers' safety and health.

References

SPRING Singapore. (2015). SS 607: 2015 Specification for design of active fall-protection systems. Singapore: SPRING Singapore.

Tranter, M. (2004). *Occupational hygiene and risk management*. 2nd edition. Sidney: Allen & Unwin.

Workplace Safety and Health Council. (2013). Code of practice for working safely at heights. 2nd revision. Singapore.

Workplace Safety and Health Council. (2014). Code of practice on safe lifting operations in the workplaces. Singapore. https://www.wshc.sg/files/wshc/upload/cms/file/2014/Code_of_Practice_Safe_Lift-ing_Operations_Revised_2014.pdf (accessed: 27 March 2018).

Workplace Safety and Health Council. (2017). Asbestos. Singapore.

CHAPTER 10

Accident case studies

10.1 Introduction

This chapter presents three accident case studies to facilitate discussions about key workplace safety and health (WSH) management concepts presented in the earlier chapters. At the same time, relevant WSH legislation will also be highlighted.

10.2 Case A: Crane collapse

10.2.1 Background

This case is taken from Jurong Primewide Pte Ltd v Moh Seng Cranes Pte Ltd and others [2014] 2 SLR 360. Referring to Fig. 10.1, Jurong Primewide Pte Ltd ("JPW") was the main contractor building a seven-storey multi-user business park development for Crescendas Bionix Pte Ltd ("Crescendas"). JPW had a crane rental agreement with Hup Hin Transport Co Pte Ltd ("Hup Hin") on a per-call basis, and Hup Hin had a hiring contract with Moh Seng Cranes Pte Ltd ("Moh Seng") to hire Moh Seng's mobile cranes whenever required. Moh Seng was the crane operator's employer, and their crane was damaged in this accident. MA Builders Pte Ltd ("MA") was JPW's subcontractor responsible for structural, architectural, and external works.

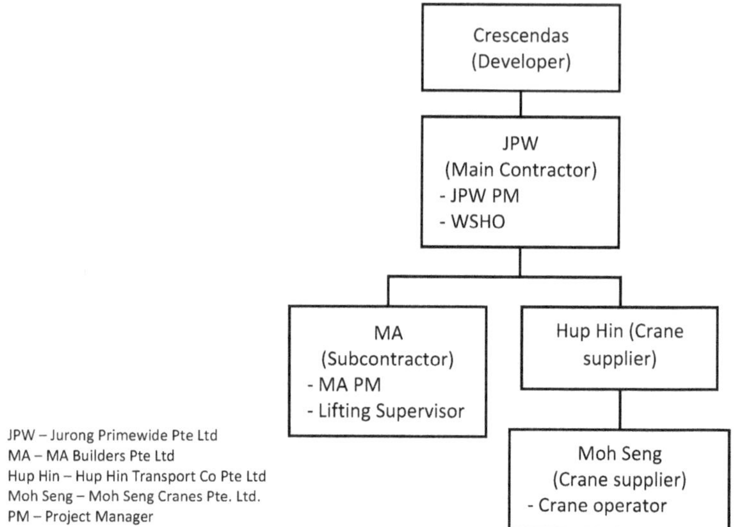

Fig. 10.1 Relationship between the different parties of Case A.

10.2.2 Accident detail

On 10 June 2010, MA requested for a mobile crane from JPW to lift rebars. Thus, JPW contacted Hup Hin to deliver a 50-tonne mobile crane to the worksite the following day. As Hup Hin did not have a 50-tonne mobile crane on 11 June 2010, they contacted Moh Seng for the crane.

On 11 June 2010, the crane operator employed by Moh Seng arrived at the site with the 50-tonne mobile crane. He reported to the lifting supervisor employed by MA, who instructed the crane operator to park at the washing bay area near the guardhouse in the vicinity of a manhole covered by soil. The crane operator expressed concerns that the designated location might not be able to take the crane's weight safely. The lifting supervisor assured the crane operator that the location had "hard flooring" and could support the crane's weight. The crane operator was unconvinced and expressed his concerns to JPW's workplace safety and health officer (WSHO). The WSHO discussed with the lifting supervisor and then reassured the crane operator that the ground could take the loading imposed by the crane. Thus, the crane operator proceeded to deploy the crane in compliance with the lifting supervisor's instructions.

When the crane operator swung the boom from the left front of the crane towards the left back outrigger, the left back outrigger broke through the cover of a concealed manhole, causing the crane to collapse. Layers of brown soil had covered the manhole, making it hidden before the accident.

It was established that JPW knew about the manhole by 12 August 2008 but did not take reasonable care to mark and cordon off the manhole properly. They also did not direct the subcontractors to take the appropriate action. JPW indicated that the manhole was usually visible and so "obvious to any person or passer-by" that there would not have been "the need of [sic.] any barricade, warning sign or warning". They also blamed MA for allowing soil to accumulate over the manhole due to their excavation work. However, the Judges found that the sheer knowledge of the presence of the manhole would have made it foreseeable that the accident could happen on a busy site with heavy equipment and machinery. Thus, JPW's negligence was inexcusable.

Nevertheless, MA, as the subcontractor, was also expected to ensure that the lifting operation was conducted safely. This is especially the case because MA was the employer of the lifting supervisor who was instructing the lifting work. Furthermore, MA was fully informed of the location of the manhole because they conducted an investigation and piping works in the manhole before the accident. MA's project manager (PM) had also warned the lifting supervisor about the manhole, and the lifting supervisor acknowledged that he was aware of the presence of the manhole. Despite this knowledge, MA's lifting supervisor commenced lifting operations without a valid permit-to-work (PTW).

10.2.3 Event Causation Technique analysis

Figures 10.2 and 10.3 show the why-analysis and Event Causation Technique (ECT) diagram for the accident, respectively. The breakdown event is, "Left back outrigger placed over concealed manhole". The direct cause of the accident was the decision made by the lifting supervisor and JPW WSHO to locate the crane near the washing bay. This decision was probably based on their assumption that the manhole cover could take the load from the outrigger and/or the outrigger was not placed directly over the concealed manhole. The lifting supervisor and WSHO could have been improperly motivated, but there was a lack of information on why the two safety personnel decided, despite concerns from the crane operator. The concealment of the manhole was also a direct cause leading to the outrigger being placed over the concealed manhole.

One control that could have been implemented to prevent the breakdown event is a PTW system to ensure that legal requirements, such as checking on the ground conditions were conducted before lifting. Reg. 12 of the Workplace Safety and Health (Construction) Regulations 2007 (No. S 663/2007) required a PTW to be first issued by the PM of the worksite before high-risk construction work could be conducted (see Chap. 9). At the time of the accident, Reg. 20(3)(c) of the Factories (Operations of Cranes) Regulations 1998 ("the Regulations") was still in force, and it stated that a lifting supervisor has to "ensure that the ground conditions are safe for any lifting operation to be performed by any mobile crane". This was also highlighted in the Singapore Standard SS 536 2008 Code of Practice for the Safe Use of Mobile Cranes, which specifically identified manholes as one of the possible "underground hazards" that must be checked for.

Another possible control that could have prevented the outrigger from being placed near or over the manhole was to

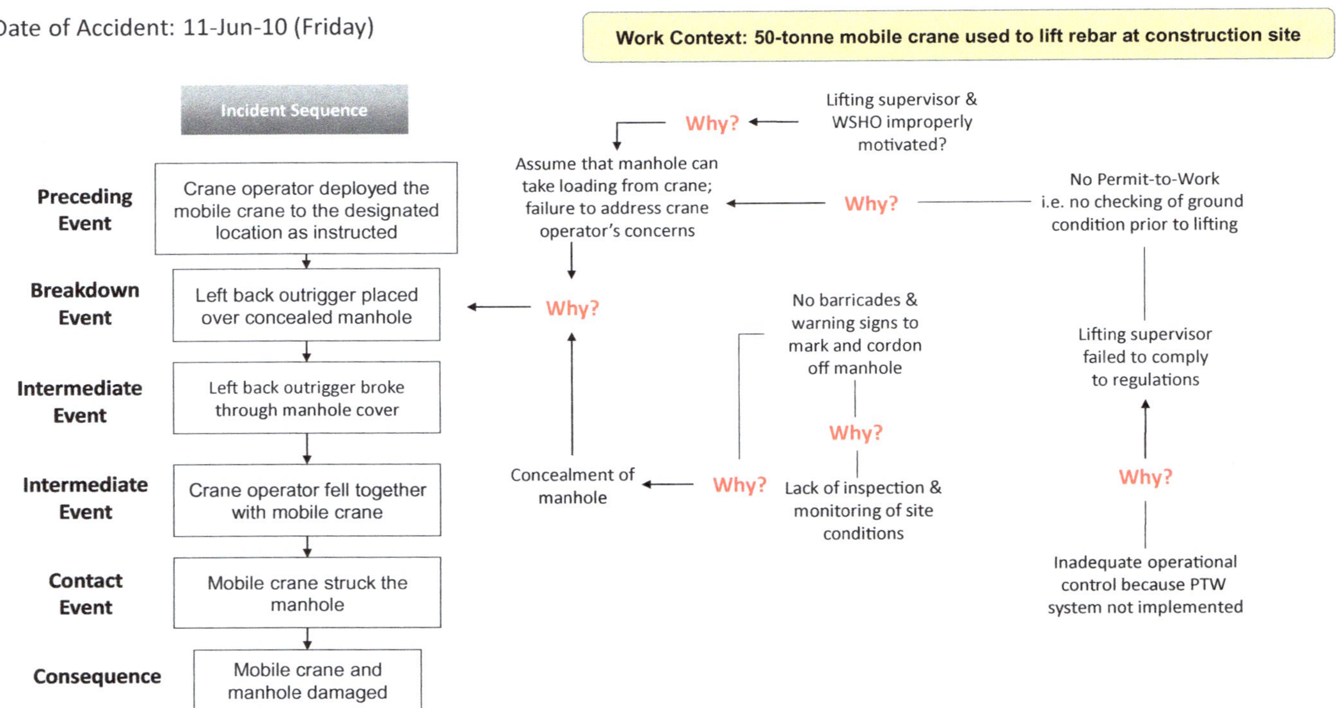

Fig. 10.2   Why-analysis for Case A.

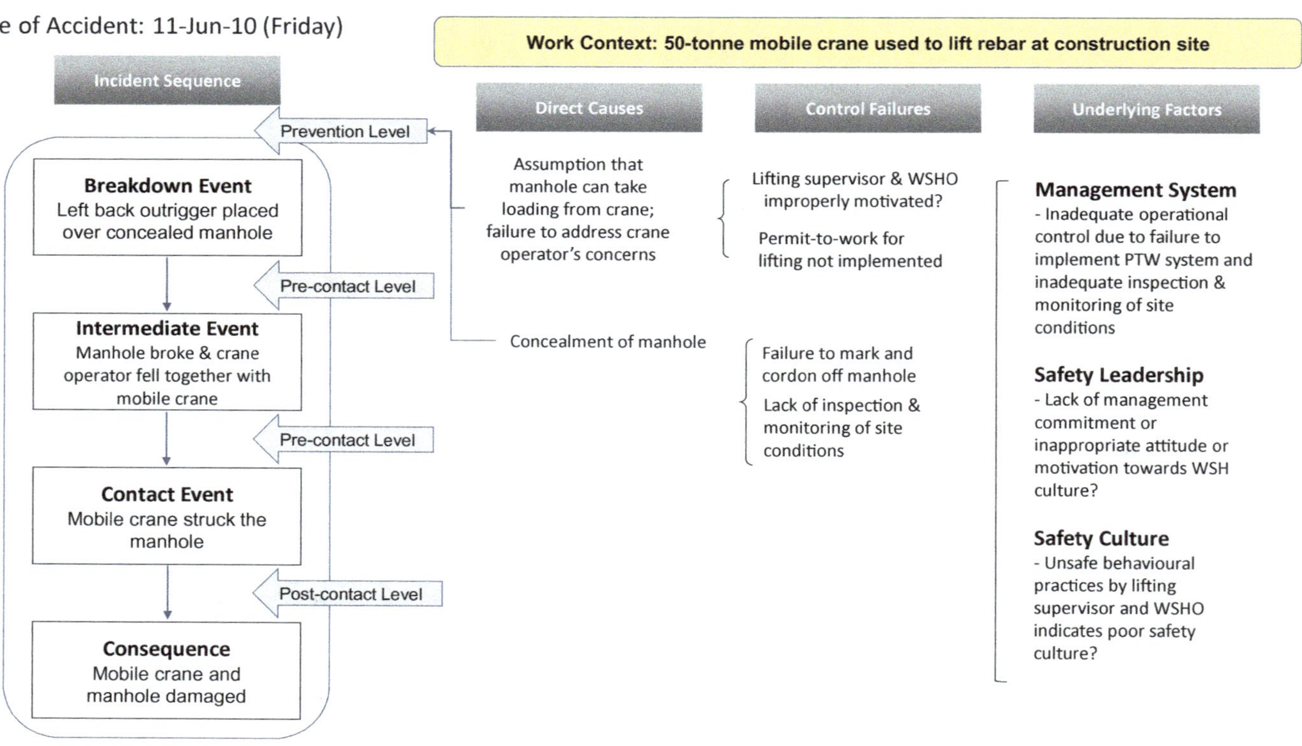

Fig. 10.3　Event Causation Technique diagram for Case A.

mark and cordon off the manhole area. This could also prevent the manhole from becoming concealed by the soil during excavation.

Through the actions of the lifting supervisor and WSHO, who ignored the concerns of the crane operator about the ground condition, it is inferred that the safety culture and leadership of the site were probably lacking before the accident. In addition, with the failure to implement the PTW system for high-risk activities like lifting and the lack of site inspection and monitoring, it can be inferred that the level of implementation of the operations control element of the management system was probably inadequate.

10.2.4 Discussion

This discussion will focus on the legislative implications of the case. WSH management system concepts will be discussed at the end of the chapter.

In their judgement, the Appeal Judges highlighted the intention of the Workplace Safety and Health Act (WSHA) to create a system of accountability by defining the duties of different stakeholders. The Minister for Manpower, Dr. Ng Eng Hen, during the second reading of the Workplace Safety and Health Bill 2005 (Bill 36 of 2005) (Singapore Parliamentary Debates, Official Report (17 January 2006) vol 80) at cols. 2208–2210, explained:

> "...this Bill will better define persons who are accountable, their responsibilities and institute penalties which reflect the true economic and social cost of risks and accidents. Penalties should be sufficient to deter risk-taking behaviour and ensure that companies are proactive in preventing incidents. Appropriately, companies and persons that show poor safety management should be penalised even if no accident has occurred. This Bill will put into place a new and more effective

framework to reduce accidents at the workplace — to bring about a quantum improvement in OSH (occupational safety and health) standards and to achieve our intermediate goal of halving the present occupational fatality rate by 2015."

Accordingly, contractors and subcontractors, i.e., JPW and MA, respectively, are two of the entities that the WSHA seeks to "increase direct liability on for workplace safety". Main contractors and subcontractors have direct "operational control" of workplaces, and they have tremendous responsibilities to ensure a safe working environment at construction sites. Even though the main contractor, as the Occupier of the work site, has a duty under Section 11 of the Workplace Safety and Health Act (Chapter 354A) to ensure, so far as is reasonably practicable, that the workplace, all means of access to or egress, and any machinery, equipment, plant, article, or substance are safe, principals (MA in this case) and employers have an important role because they direct the manner of work. Dr. Ng Eng Hen, in the second reading of the Workplace Safety and Health Bill 2005, also explained how the WSHA sought to impose duties on principals engaging contractors for specialised tasks (at col. 2209):

"Traditionally, a principal who engages a contractor would be engaging the specialist services of the contractor, and would not be directing the contractor on how to do the work. However, today the situation is different. Principals often engage 'contractors' and third-party labour not for their specialist expertise, but precisely so that they can avoid entering into a direct employment relationship, for organisational or other reasons. In such situations, the principal, in terms of supervision, takes on the role of an employer. The Bill thus places on him responsibility for the worker's safety and health as if he were the employer. If this were not the case, then the duties under the Act could be simply circumvented by a careful crafting of the legal relationship."

Thus, JPW, as the main contractor, had to ensure that "hazards such as manholes were identified properly through site surveys, and then to undertake ground improvements to ensure that these underground hazards did not continue to remain hazardous". Even though MA was "not contributorily negligent in causing some soil run-off to cover the manhole due to its nearby excavation works", it is still responsible for the accident because the "lifting operation that caused the accident was within MA's scope of work" (the lifting supervisor, riggers and signalmen were employed by MA), "MA knew about the manhole as well", and "MA was contractually responsible and deemed fully informed of the conditions at the worksite".

The lifting supervisor is an important safety personnel during a lifting operation, but in this case, the lifting supervisor failed to perform his duties specified in Reg. 20(3)(c) of the Factories (Operations of Cranes) Regulations 1998, which states that a lifting supervisor has to "ensure that the ground conditions are safe for any lifting operation to be performed by any mobile crane". Furthermore, the Singapore Standard SS 536 2008 Code of Practice for the Safe Use of Mobile Cranes required the lifting supervisor to (italicised text are relevant duties for the case study):

(a) co-ordinate and be present to supervise all lifting activities and ensure that the lifting operation is carried out safely;
(b) ensure that only registered crane operators, appointed riggers, and appointed signalmen participate in any lifting operation involving the use of a mobile crane;
(c) ensure that the ground conditions are safe for any lifting operation to be performed by any mobile crane;
(d) ensure that there is a set of safe lifting procedures for any lifting operation of a mobile crane;
(e) brief all crane operators, riggers and signalmen on the safe lifting procedures referred to in (d);

(f)   take measures to rectify the unsatisfactory or unsafe conditions that are reported by any crane operator or rigger.

The Judges were especially critical of the lifting supervisor because, though he was aware that there was a concealed manhole, he failed to ensure that the hazard was effectively managed. Worse still, the lifting supervisor failed to take any measures to address the crane operator's concerns about the safety of the ground conditions. Thus, the Judges "had no hesitation in finding that MA was liable in negligence for the damage caused during the accident".

Finally, the Judges made the following decision: "[B]oth JPW and MA were liable in negligence to Moh Seng, and that MA did breach the subcontracts that it had entered into with JPW, we felt that an apportionment of 60% to JPW and 40% to MA would be just in all the circumstances. As the main contractor, as well as the occupier under the WSHA, JPW had to bear the bulk of the responsibility for failing to identify an underground hazard like the manhole and for not taking measures to ensure that it ceased to be an unknown danger. However, we were also cognisant of the fact that MA was largely in charge of the works that were in the area, and specifically, the operation of lifting steel rebars that [the crane operator's] crane was engaged in. Also, the Lifting Supervisor was under MA's employment and had specific duties under the WSH Regime which he failed to fulfil."

10.3 Case B: Falling from loading platform during lifting

10.3.1 Background

This case is taken from Public Prosecutor v GS Engineering & Construction Corp [2016] SGHC 276. GS Engineering & Construction

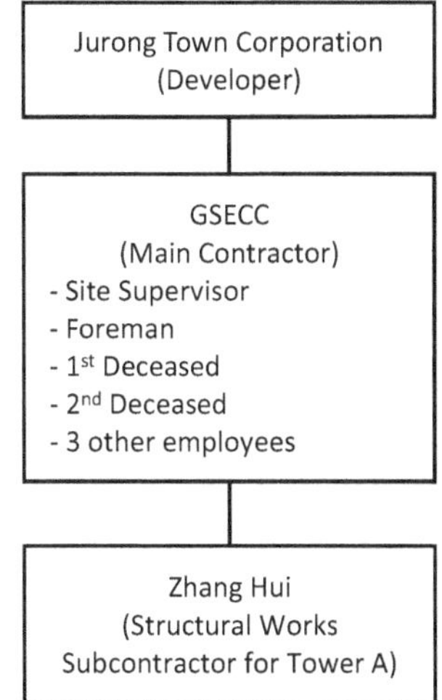

Fig. 10.4   Parties involved in Case B.

Corp (GSECC) was the main contractor to construct two towers (Tower A and Tower B) for Jurong Town Corporation (Fig. 10.4). The two towers were to be 11 and 18 storeys high, respectively. GSECC subcontracted the structural works (including formwork) for Tower A to Zhang Hui Construction Pte Ltd ("Zhang Hui"). The project started on 23 November 2011 and was meant to be completed by 23 March 2014.

10.3.2 Accident detail

On 22 January 2014, GSECC employees were initially planning to shift the loading platform from the tenth floor of Tower B to the eighth floor of Tower A. Loading platforms are typically positioned on the edge of buildings to facilitate the lifting of materials and items to and from different levels of the building and the ground level. When employees from Zhang Hui requested for the GSECC site supervisor to move the air compressor from the seventh to the eighth floor of Tower A using the loading platform, the site supervisor instructed the GSECC foreman to assist Zhang Hui with this task. The site supervisor also instructed the foreman "not to install the loading platform at the seventh storey of Tower A, but to simply suspend it by a tower crane". It is presumed that the site supervisor wanted to save time so that the time taken for collecting the air compressor at level seven could be minimised.

At about 11.50 am, the foreman with five workers, including the two deceased, shifted the loading platform from Tower B to the seventh storey of Tower A. As Zhang Hui's workers were at lunch, the GSECC foreman proceeded with his workers. The air compressor was mounted on a steel frame with four wheels, but the rear wheels were smaller and could not roll onto the suspended loading platform. In addition, during the loading process, the suspended platform tilted. One of the deceased workers and his co-workers informed the foreman that pushing the air compressor onto the suspended platform was unsafe. However, the foreman instructed them to continue with the task.

The workers then used a galvanised pipe to pivot the air compressor onto the loading platform. After several attempts, the air compressor was pushed onto the platform, but it rolled towards the two deceased on the platform, causing it to tilt. The two deceased could not evade the air compressor in time and fell together with it.

The air compressor landed on another loading platform two levels below. The two deceased fell to the ground level and were pronounced dead by the paramedics attending to the accident.

10.3.3 Event Causation Technique analysis

The why-analysis (Fig. 10.5) and ECT diagram (Fig. 10.6) summarise the incident causation. The breakdown event is "Platform tilted & air compressor rolled towards 2 deceased", and the contact event is "2 deceased hit the ground".

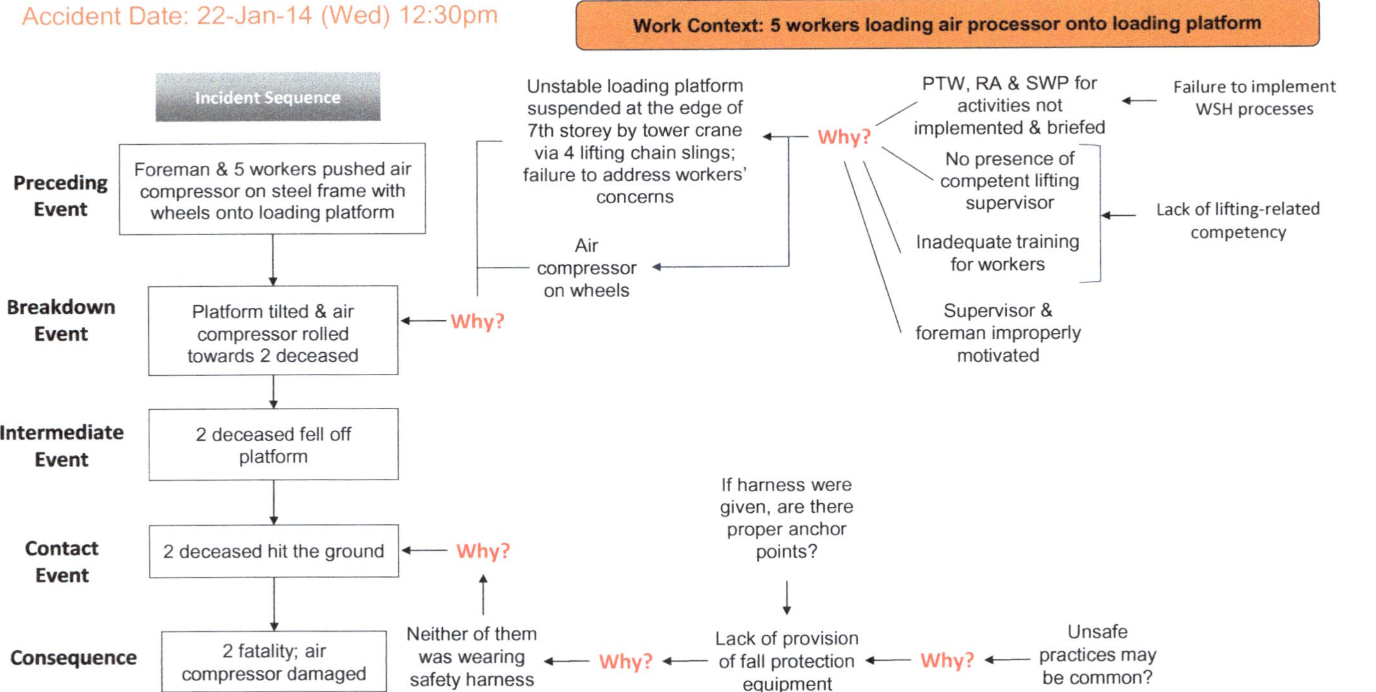

Fig. 10.5   Why-analysis for Case B.

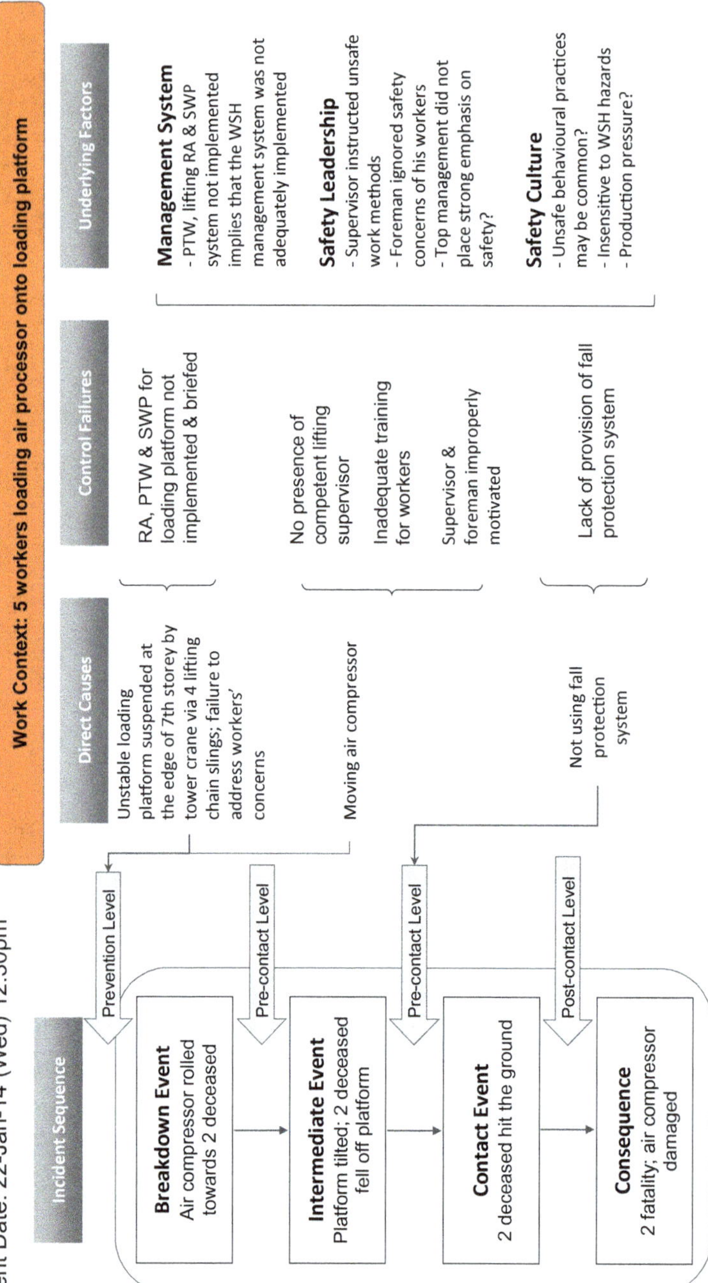

Fig. 10.6   ECT diagram for Case B.

The breakdown event happened because, before it, the workers were pushing the air compressor on wheels onto the platform and the loading platform was suspended by the tower crane using four lifting chain slings. Furthermore, the foreman failed to address the workers' concerns, who felt that it was unsafe to proceed with the work because the platform tilted when they were loading the compressor onto the platform. Effective training on the safe use of the loading platform, the presence of a competent lifting supervisor, the implementation and briefing of the PTW, and risk assessment (RA) and safe work procedure (SWP) for the lifting and use of the loading platform should have resulted in the removal of the unsafe work method and hence prevented the breakdown event. Even though a foreman was present, he was likely overly focused on completing the task and failed to ensure the workers' safety. Similarly, the supervisor had probably instructed the unsafe work method to save time. In addition, the lack of a fall protection system (e.g., fall arrest anchor, lifeline and harness) for the workers on the day of the accident meant that it was very unlikely to have prevented the fatality when the breakdown event happened.

There is a lack of information on the state of the safety culture of the project, but based on the actions of the foreman and supervisor, it seems that frontline leaders (supervisors and foremen) may not have placed due emphasis on safety. Similarly, several operational controls like the PTW, RA and SWP not being adequately implemented are signs that the management system is ineffective, indicating inadequacies in the organisation's safety culture and safety leadership.

10.3.4 Discussion

This discussion will focus on the legal implications of the case. WSH management implications will be discussed at the end of the chapter.

This is a landmark case because it was the first time a WSHA case was presented to the High Court. Furthermore, the Prosecution disputed the District Judge's sentencing and proposed a new sentencing framework. One of the key rationales for the Prosecution's proposal was that the sentencing for previous cases was too low, with the majority falling below 30% of the maximum sentence of the S$500,000 maximum fine prescribed in the WSHA. The relatively low sentencing was contrary to the legislative intent of the WSHA, which then Minister for Manpower, Dr. Ng Eng Hen, described at the second reading of the Workplace Safety and Health Bill (Bill 36 of 2005) ("the Bill") as follows:

> "Three fundamental reforms in this Bill will improve safety at the workplace. First, the Bill will strengthen proactive measures. Instead of reacting to accidents after they occur, which is often too little too late, we should reduce risks to prevent them. To achieve this, all employers will be required to conduct comprehensive risk assessments for all work processes and provide detailed plans to minimise or eliminate risks.
>
> Second, industry must take ownership of occupational safety and health standards and outcomes to effect a cultural change of respect for life and livelihoods at the workplace...
>
> Third, this Bill will better define persons who are accountable, their responsibilities and institute penalties which reflect the true economic and social cost of risks and accidents. Penalties should be sufficient to deter risk-taking behaviour and ensure that companies are proactive in preventing accidents. Appropriately, companies and persons that show poor safety management should be penalised even if no accident has occurred."

Furthermore, Dr. Ng explained the reason for the need for higher penalties for poor safety management and performance:

> "Even as we work with industry to build up their capabilities to improve safety and health at their workplaces, we need to ensure that the penalties for non-compliance are sufficiently

high to effect a cultural change on the ground. Penalties should be set at a level that reflects the true cost of poor safety management, including the cost of disruptions and inconvenience to members of the public which workplace accidents will cause. The collapse of the Nicoll Highway not only resulted in the loss of four lives, but also caused millions of dollars in property damage and led to countless lost working hours and great inconvenience to the public. The maximum penalty of [S]$200,000 under the present Factories Act is therefore inadequate."

The High Court Judge adopted a sentencing framework that was further discussed and updated in subsequent cases. Of particular interest are the MW Group Pte Ltd v Public Prosecutor [2019] 3 SLR 1300 ("MW Group"), Mao Xuezhong v Public Prosecutor and another appeal [2020] 5 SLR 580 ("Mao Xuezhong"), and Public Prosecutor v Manta Equipment (S) Pte Ltd [2022] SGHC 157 ("Manta Equipment"). In Manta Equipment, the latest of the series of cases that updated the sentencing framework, the Judges supported the two-stage sentencing framework developed in the GSECC case and updated the sentencing framework (Table 10.1). As observed in Table 10.1, a court will determine the appropriate starting point for the sentence by considering two principal factors: (i) the culpability of the offender and (ii) the level of harm, including the potential harm and the actual harm, as described below:

1. Determine the culpability of the offender based on the following non-exhaustive factors:
    a. the number of breaches or failures in the case;
    b. the nature of the breaches;
    c. the seriousness of the breaches — whether they were a minor departure from the established procedure or whether they were a complete disregard of the procedures;
    d. whether the breaches were systemic or whether they were part of an isolated incident;
    e. whether the breaches were intentional, rash or negligent.

Table 10.1   Sentencing framework determined by High Court (all monetary figures shown are in SGD).

		Culpability		
		Low	**Moderate**	**High**
Harm	High	$150,000 to $225,000	$225,000 to $300,000	$300,000 to $500,000
	Moderate	$75,000 to $150,000	$150,000 to $225,000	$225,000 to $300,000
	Low	Up to $75,000	$75,000 to $150,000	$150,000 to $225,000

2. The level of harm may be assessed by considering, among other things, (i) the seriousness of the harm risked, (ii) the likelihood of that harm arising, and (iii) the number of people likely to be exposed to the risk of the harm. If the harm was likely to be death or serious injury (e.g., paralysis or loss of a limb), it could be considered high even though it did not materialise. Death and serious injury would be considered a high level of harm.

3. After deriving the starting point for sentencing, the court should calibrate the sentence by taking into account the aggravating factors and mitigating factors of the case.

4. Aggravating factors include the following: (i) serious actual harm (including death) resulted; (ii) the breach was a significant cause of the harm that resulted — in this regard, a significant cause need not be the sole or principal cause of the harm, and need only be a cause that has more than minimally, negligibly, or trivially contributed to the outcome; (iii) the offender had cut costs at the expense of the safety of the workers; (iv) there was a deliberate concealment of the illegal nature of the activity; (v) there was a breach of a court order; (vi) there was an obstruction of justice; (vii) the offender has a poor record with respect to workplace health and safety; (viii) there was falsification of documentation or licences, and (ix) there was a deliberate failure to obtain

or comply with relevant licences in order to avoid scrutiny by the authorities.

5. Mitigating factors may include the following: (i) the offender has voluntarily taken steps to remedy the problem; (ii) the offender provided a high level of cooperation with the authorities for the investigations beyond that which is normally expected; (iii) there is self-reporting, cooperation, and acceptance of responsibility; (iv) there is a timely plea of guilt; (v) the offender has a good health and safety record, and (vi) the offender has effective health and safety procedures in place.

For the case of GSECC v PP, the Judge opined that GSECC's (Respondent) culpability falls into the Medium to High category because the occupier is overall in charge of the worksite. Concurrently, GSECC was also the employer of the two deceased and the other workers involved in the task that resulted in the accident. Furthermore, the occupier failed to perform numerous control measures that are expected of their own lifting operations, and it also failed to ensure that its workers were adequately trained and had adequate fall protection. Thus, even if Zhang Hui had not asked GSECC to shift the air compressor, the occupier would have committed numerous breaches and failures. Moreover, the occupier oversees the PTW system and must evaluate and approve Zhang Hui's application to conduct the lifting operation.

During the trial, the Respondent attempted to shift the responsibility to its workers, but the Judge noted Dr. Ng Eng Hen's observations at the second reading of the Workplace Safety and Health Bill (Singapore Parliamentary Debates, Official Report (17 January 2006) vol. 80 at col. 2205):

"The reality is that on a day-to-day basis, safety may be the last thing on the minds of management and workers on the ground. There are deadlines to meet, monotony, apathy or lethargy to overcome, a lack of professionalism and training, unclear lines or no lines of accountability, and poor management…"

Even though workers have duties under the WSHA to cooperate with their employers, the intent of the WSHA is for the employer, occupier, or other responsible persons to "ensure that its workers are trained and are mindful of their safety at the workplace and that proper systems are in place to ensure that steps are taken to minimise risks".

Regarding potential harm, the Judge selected the "high" category because the high-risk work was conducted without a safe system of work and because fall arrest equipment was not provided. Furthermore, many of the workers were not trained for the task. Thus, all the workers involved could have been killed. In addition, the landing platform could have failed and landed on other workers.

Therefore, based on the level of culpability and the potential for harm, the starting sentence was a fine of S$300,000.

The Judge then considered the aggravating and mitigating factors to calibrate the sentence. Aggravating factors included two lives being lost as a result of the breaches. Mitigating factors included the fact that the Respondent had pleaded guilty, cooperated with the authorities, had a good safety record, won past awards from the Land Transport Authority and the Ministry of Manpower (two WSH awards for the accident site), and had been proactive in investigating the accident and implementing remedial actions. Finally, after weighing all the relevant factors, the Judge arrived at an appropriate sentence of S$250,000.

The sentencing framework is also relevant to the concept of just culture discussed in Chap. 7. When organisations establish accountabilities and responsibilities, they must also decide on the punishment for non-compliance. The punishment can be modified from the sentencing framework and tailored to the organisational context.

10.4 Case C: Falling from mezzanine floor during welding work

10.4.1 Background

This case is based on a consultancy project the author conducted in 2018. The client has agreed to allow the author to discuss the accident, but it must be clarified that this case study is based on the author's opinions and interpretations. Fictitious names were used to prevent any possible complications.

Figure 10.7 shows the relationship between the key parties involved in this accident case study. The developer is a public agency, and they selected the main contractor to construct a two-storey multi-user motor workshop. The main contractor subcontracted the fabrication, supply and installation of metal works to the subcontractor. The deceased and his two co-workers were employed by a labour supplier who supplies

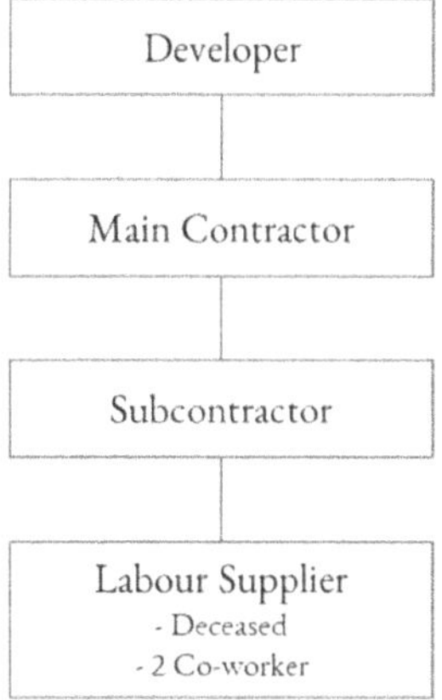

Fig. 10.7   Parties involved in Case C.

short-term workers to different construction worksites and other workplaces. In this case, the deceased and his co-workers worked for the metal work subcontractor. The deceased was a 45-year-old male Chinese worker from China.

This case study focuses on the duties of the subcontractor, who is deemed the principal controlling and directing the work of the deceased and his co-workers and hence had the same duties as an employer under the WSHA. The project commenced on 2 April 2012, and the subcontractor's contract with the main contractor was made on 25 December 2012. The subcontractor was engaged to "supply and install the galvanised steel hollow section with steel plates and bolts to the mezzanine office dry wall partitions at the 1st and 2nd storeys". A mezzanine floor is an intermediate floor in a unit of a building (e.g., an apartment or a motor workshop) that does not cover the whole floor space of the unit.

10.4.2 Accident detail

On 27 April 2014, the deceased and his two co-workers were installing steel hollow sections (SHSs) on the mezzanine floor in a second-level motor workshop unit under construction (see Fig. 10.8). The SHSs were used to construct the drywall's metal frame for the mezzanine floor. Each metal frame for a drywall was mainly made up of vertical columns and two horizontal beams (upper and lower SHS). The mezzanine floor was 2.8 m above the floor of the unit. For the deceased, it was his first day on this site.

Before the accident, the deceased (worker 1 in Fig. 10.8) was hammering the upper SHS (about 5.1 m long) to level it while standing on the lower SHS and/or the edge barricades with his co-worker (worker 2). Even though there were discrepancies about the position of the deceased and his co-worker, based on the case details, the author opined that it was most

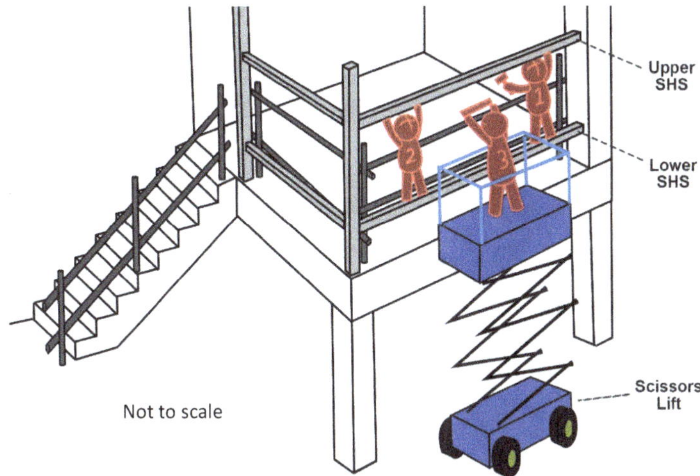

Fig. 10.8   Overview of work prior to accident.

likely that they were indeed standing on the lower SHS and/or the barricades. The other co-worker (worker 3) was working from a scissors lift (a type of mobile elevated work platform (MEWP)) placed next to the mezzanine floor. The lower SHS was tack-welded into place. It is noted that tack welds are meant for preliminary positioning and alignment purposes, and they are not meant to take significant load. Furthermore, the barricades were not properly secured and were simply placed near the edge.

At about 3.45 pm, the deceased fell from the mezzanine floor and was conveyed to the hospital, where he succumbed to his injuries on 2 May 2014 at 3.45 am.

10.4.3 Event Causation Technique analysis

The why-analysis (Fig. 10.9) and the ECT diagram (Fig. 10.10) summarise the incident. The breakdown event could not be definitively confirmed, but the likely causes were "tack welded lower SHS failed" and "unsecured barricade shifted". When the

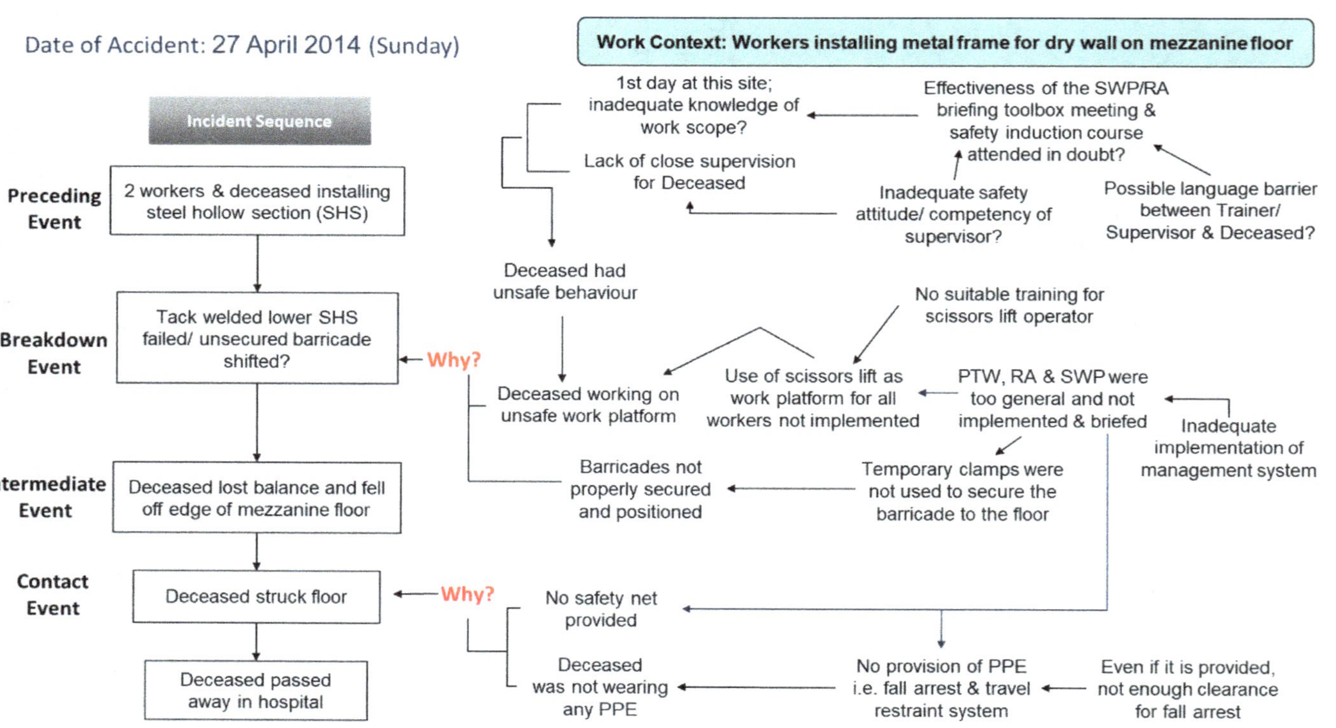

Fig. 10.9   Why-analysis for Case C.

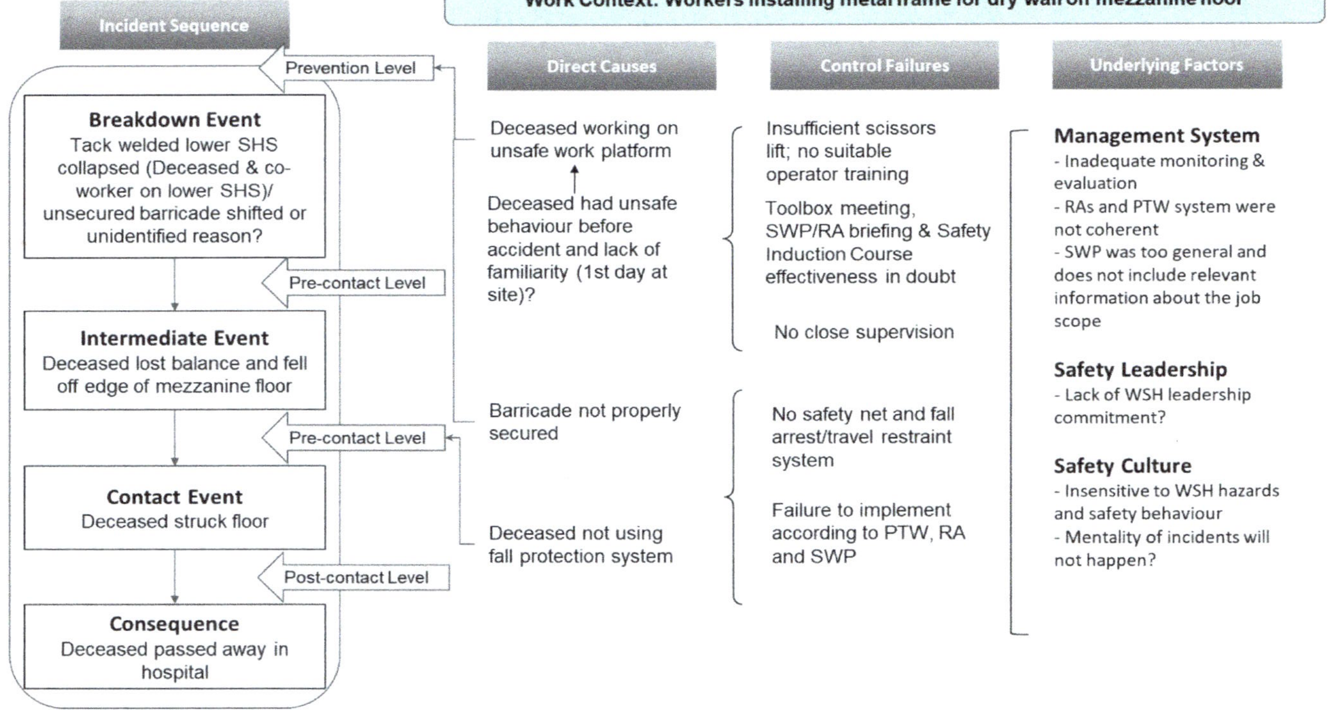

Fig. 10.10   ECT diagram for Case C.

lower SHS and/or the barricade moved, the deceased lost balance and fell onto the floor. The two main incident events that should be evaluated are the breakdown and contact events.

Before the accident, the deceased was standing on an unsafe work platform (the tack welded lower SHS and/or the unsecured barricades). It must be noted that the workers were not provided with the necessary safe working platform. According to a set of work procedures that the subcontractor produced, the deceased and his co-workers were supposed to work from two scissors lifts, but they were only provided one scissors lift on the day of the accident, which was not enough for them to do their work properly. Furthermore, the deceased and his co-workers were not trained to operate the scissors lift they used. The deceased was also not trained for welding and working at height (WAH).

The barricades were not properly secured and positioned. Hence, it was unsafe even if the workers were expected to use the mezzanine floor to conduct the work. Furthermore, the workers were not provided with a suitable step ladder or platform to reach the height of the upper SHS. Thus, the workers had to work unsafely to position themselves to conduct the welding work.

The subcontractor felt that the workers' poor safety attitude led to the accident. The deceased was issued an "EHS Administrative Charge" (essentially a site-imposed fine) on the day of the accident. The main contractor found him to be working unsafely at height without proper personal protective equipment (PPE). The main contractor instructed the subcontractor to stop the work-at-heights. The subcontractor supervisor received the "EHS Administrative Charge".

The unsafe behaviour of the deceased could be related to the fact that it was his first day working at the site, and he might have been unfamiliar with the site requirements and work conditions. His safety attitude and behaviour could have been managed through briefings, training, and close supervision. However, these control measures were ineffective.

The site had a series of briefings and training for the deceased and his co-workers when they arrived at about 8 am on the day of the accident. These included orientation, a safety induction course, SWP and RA briefing, and a toolbox meeting. The deceased also signed two PTW forms. All these activities happened between 8 am and 9 am. It was also noted that the supervisor was from India, while the deceased and co-workers were from China. Based on the author's experience, the language barrier between workers of different nationalities was (and continues to be) a major challenge in the Singapore construction industry. Thus, given the language barrier and limited time, it was very unlikely that the training and briefings were in-depth and effective. However, the contributions of language barrier to the accident was not established.

Furthermore, knowing that the deceased was working on the site for the first time, the subcontractor should have supervised his work closely. Close supervision means that the supervisor must be able to intervene promptly when the deceased is conducting his work. Close supervision was even more important after the deceased received the "EHS Administrative Charge" but was allowed to continue to work on high-risk tasks such as work-at-height and welding. However, the supervisor did not closely supervise the deceased and his co-workers. If close supervision was not possible, the alternative was to disallow the deceased to continue with the high-risk work. The lack of actions by the supervisor cast doubts over his safety attitude and competency.

Regarding the contact event, where the deceased struck the floor after falling off the mezzanine floor, a relevant control would have been a fall protection system. A well-designed travel restraint system, safety net, or fall arrest system can pos-

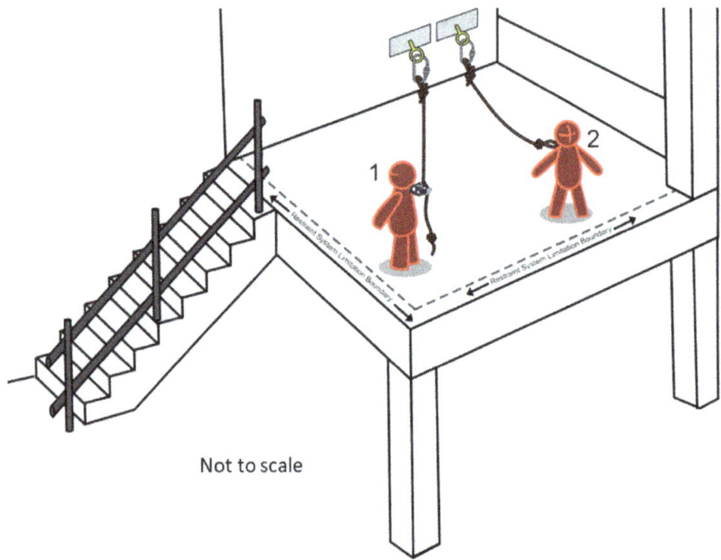

Fig. 10.11   Possible travel restraint systems to protect the workers.

sibly have prevented the fatality. Figure 10.11 shows two possible alternatives for implementing a travel restraint system, which the subcontractor and the main contractor could have been provided with prior to the accident. Option 1 in Fig. 10.11 is an adjustable travel restraint system, and option 2 is a fixed length travel restraint system. If a fall arrest system was used, the lack of fall clearance would probably mean that a self-retracting lanyard had to be used. Even though the workers were provided with full body harnesses and lanyards, they did not have the proper anchors to protect themselves, and the lanyards were not suitable because the extension of the personal energy absorber during a fall arrest would mean that the clearance of 2.8 m was inadequate.

The subcontractor had a suite of documents, albeit at times incoherent, which contained many of the control measures discussed above. For example, the PTW indicated that a fall arrest system, including a lifeline, should be provided, but no fall arrest anchors or lifelines were identified or installed. It was also noted that the project manager had signed off on the PTW, but the WSHO did not sign off even though it was required in the form. The PTW also indicated requirements for a fire extinguisher, flashback arrestor, cylinder, hoses, "O" clips, and leakage checks to be provided, but no fire extinguisher could be observed, and the workers were using an electrical welding set, i.e., the flashback arrestor, cylinder, hoses, "O" clips, and leakage checks were not relevant.

The RAs and SWPs were also in a mess. Many of the safety measures were general and not specific to the work and were not applicable. Linkages between the documents were not established. They appeared to be independent of each other and were of inadequate quality. The RAs were very general and not specific to the workers' activities. They also had mistakes in their calculations of risk levels.

The wide range of control failures and inadequacies in the RA, PTW system, and SWPs showed that the WSH management system was inadequately implemented. With the apparent insensitivity to hazards and the unsafe behaviour of the workers and supervisor, it is justified to assess that the safety culture was very poor. Although no information on the top management's level of commitment to WSH was provided, based on the ineffective WSH management system and poor safety culture, it is hard to imagine that the top management of the subcontractor is dedicated to WSH.

10.4.4 Discussion

The case demonstrated a severe disregard for safety by the subcontractor. The actual manner of work of the deceased and his co-worker were very dangerous because they were not on any safe work platform and were not protected by any fall protection system or PPE. The work procedure did not follow applicable safety codes, standards and guidelines referred to in the evaluation.

The deceased had displayed unsafe behaviour before the accident, was not trained for the welding work nor WAH, and was unfamiliar with the work site. Yet, he was conducting two high-risk tasks concurrently. Thus, close supervision was expected but was not implemented. The supervisor might not have been able to communicate clearly with the deceased. The unsafe behaviour that the deceased displayed may be induced by the lack of proper safety and work provisions (e.g., a raised platform for him to reach the upper SHS). The situation in this case could have induced a "necessary violation" (Reason, 1997, p. 73), where

> "[N]on-compliance is seen as essential in order to get the job done. Necessary violations are commonly provoked by organizational failings with regard to the site, tools or equipment… In addition, they can also provide an easier way of working. The combined effect of these two factors often leads to these violations becoming routine rather than exceptional."

The deceased and his co-workers were expected to complete their work without the second scissors lift and safe work platform. In such a situation, they may have thought it necessary to violate WSH rules to get their work done.

The effectiveness of the toolbox meeting, briefings, orientation, and safety induction course was doubtful because it was impossible for so many briefings and meetings to be conducted within one hour. It was also noted that the deceased was not trained for welding and WAH. His co-workers were also not trained to operate the scissors lift.

The range of WSH documents (RA, SWP, method statement, and PTW) were not specific to the tasks the deceased and his co-workers were conducting and the location he and his co-workers were in. The controls identified were too general, and the documents had numerous mistakes. Many of the controls identified were also not implemented. Overall, the WSH-related documents were not coherent and of low quality. The WSH management system was ineffective as it was not conducted with due care. The documents reflected a low emphasis on WSH and poor safety leadership.

The subcontractor seemed to be focused on having documents to satisfy regulatory requirements but had no intention to satisfy the spirit of the WSHA. Such is what the author would call a "paper exercise", or surface compliance, which was discussed in earlier chapters. These WSH documents were meant to facilitate WSH processes. Writing the information down requires the site personnel to consider hazards, risks and controls more deliberately and in more detail. However, in practice, these WSH processes are frequently seen as additional tasks, and some companies look for "shortcuts" to complete them.

10.5 Discussion on workplace safety and health management lessons

The three accident cases presented important WSH management lessons. These cases highlight the importance of frontline leadership by supervisors, foremen, and site personnel. These supervisory personnel make many decisions on the ground that can save or kill workers, and their decisions will be based on their perception of what is important and what they believe is the right way to conduct their work. Applying the concepts discussed in earlier chapters to ensure that supervisory personnel adhere to WSH procedures, managers must impress upon them the importance of WSH. To do so, managers must have dedicated WSH communication sessions with workers and supervisors.

At the same time, managers must "give meaning" to the administrative controls such as the PTW system, RAs and SWPs, which, as observed in the three cases discussed in this chapter, can be easily defeated and violated. As discussed in Case C, it is a common problem in workplaces that paperwork is treated as just paperwork, i.e., the controls identified are not implemented and communicated to the workers. Only when managers review and discuss the documents with site personnel will they see the relevance of these documents. If the documents are used and referred to, the quality of RAs and control measures will be improved.

Supervisory personnel must be trained on soft skills like communication and team leadership to effectively tap into the knowledge and wisdom of the workers under their charge to prevent accidents effectively. As seen from the first two cases, workers on the ground can foresee possible accidents, and failure to listen to workers' concerns can result in accidents. Supervisory personnel should address all safety concerns through the RA process. When a worker identifies a hazard, the work team should discuss the potential consequences and their likelihood in detail and systematically so that suitable measures can be implemented to prevent accidents and ill health. Such onsite and quick RAs force the supervisor or foreman to take a systematic and detailed look at the task instead of addressing the hazards intuitively, during which they may be prone to cognitive biases such as over-confidence.

Case C showed that although workers play an important role in preventing accidents and ill health, management and supervisors must still implement various measures to manage these unsafe behaviours. The WSHA, aligned with WSH management principles, clearly stipulated that workers must cooperate with their employer or principal and the occupier. Thus, when workers disregard WSH measures and the management's best intention to prevent accidents and ill health, the workers must be removed from the workplace or only allowed to work on low-risk tasks. Such drastic actions must be taken to prevent routine violations from developing in the workplace.

The range of possible hazards and accidents is very wide. Thus, when evaluating and learning from accident cases, it is important to consider the underlying factors, which are more fundamental and applicable to most workplaces. However, most investigations do not have the resources to probe the management system, culture, and leadership issues in sufficient depth. The three case studies show that only subjective inferences can be made. There is a need for more thorough investigations, and the findings must be shared widely to facilitate learning and prevention of similar incidents.

Review questions

1. If you are to develop a non-compliance penalty framework for your company, how would you use the WSH Act sentencing framework to guide the framework?
2. Based on the case studies, discuss the importance of front-line WSH leadership.
3. What are the possible ways to influence supervisors' safety behaviour?
4. Do you think the individual workers should be punished in the three case studies? Why?
5. Concerning Cases A and B, how can we encourage workers to refuse to do dangerous work?
6. How would you apply the concepts in the earlier chapters to change the culture of the organisations in the case studies?

Reference

Reason, J. (1997). *Managing the risks of organizational accidents.* Aldershot: Ashgate.

Index